Making *Millions* in Telecommunications

Also by Morton I. Hamburg

ALL ABOUT CABLE—LEGAL AND BUSINESS ASPECTS OF
CABLE AND PAY TELEVISION

Making *Millions* in Telecommunications

Morton I. Hamburg

RAWSON ASSOCIATES : New York

For Joan, Elizabeth, and John,
who are everything to me

The list of Venture Capital Firms on p. 156 is reprinted from the
June, 1984 issue of VENTURE, The Magazine for
Entrepreneurs, by special permission. Copyright © 1984 Venture
Magazine Inc., 521 Fifth Avenue, New York, NY 10175

Library of Congress Cataloging in Publication Data

Hamburg, Morton I.
  Making millions in telecommunications.

  Includes index.
  1. Telecommunication—United States. 2. Investments—
United States. 3. Stocks—United States. I. Title.
HE7775.H35   1985   332.6'722   84-42931
ISBN 0-89256-279-X

Published simultaneously in Canada by McClelland and Stewart Ltd.
Composition by Folio Graphics Co. Inc.
Printed and bound by Fairfield Graphics, Fairfield, Pennsylvania.
Designed by Jacques Chazaud
Produced by Rapid Transcript
First Edition

Contents

Acknowledgments

Many years of learning and experience go into the writing of any book, and I acknowledge the help of many people. My deep appreciation goes to all of those who have been my clients during twenty-seven years of law practice (nineteen of which were spent in communications law) and to the many brilliant students in my classes at the New York University School of Law for letting me learn at their expense. Their confidentiality has been preserved, but they know they have my thanks for the knowledge that they shared with me, especially concerning telecommunications.

I thank my law partners, past and present, for their tolerance as I expended time on other than strictly legal matters. I also thank those who have become my business associates, particularly John A. Gambling, the radio host, whose voice thousands of listeners in the New York metropolitan area know and enjoy every day, and S. Rodgers Benjamin, a very smart businessman who has learned a little bit about telecommunications of late. Both have showed me that you can make a lot of money and still be a nice guy, and I hope that I'll be able to follow their example someday.

My appreciation goes to Jim Duncan, for letting me use material from his fact-filled books on radio; to *Broadcasting*, for all of its informative stories and data; to Bob Titsch of Titsch Communications, for his *CableVision* magazine materials; and to Arthur Lipper III, for his *Venture* magazine data on venture capital firms interested in the telecommunications industry. I'm also indebted to Phil Zimmerman of Paneth, Haber & Zimmerman, for his helpful thoughts on accountants; to Bob Duffy of the Christal Company, for his constructive thoughts on sales representatives; to Cecil Richards, for his help on broadcasting brokers. My thanks also go to Ken Cox of MCI Communications; Giraud Chester of Goodson-Todman Productions; Ed Bleier of Warner Brothers Television and also president of the International Radio and Television Society; Marty Pompadur of Television Station Partners; Terry Elkes of Viacom International, Inc.; Gus Hauser of Hauser Communications; Bill Suter of Merrill, Lynch; John Reidy of Drexel, Burnham, Lambert; and Enrique Senior of Allen & Company, for sharing valuable thoughts about our industry.

I am grateful to Henry Goldberg of the law firm of Goldberg & Spector and to Theodore Pierson, Jr., of the law firm of Pierson, Ball & Dowd of Washington, D.C., for spending time and supplying me with materials on the new technology and some of the regulatory aspects of the industry.

Summer associate and researcher Audrey Rabinowitz assisted me greatly; I'm sure she will one day be an outstanding lawyer. Janet Ninivaggi, my secretary, managed to print out an entire manuscript and keep smiling; she has my continuing thanks for all that she does for me in all of my efforts.

Eleanor Rawson is an outstanding publisher; Maxine Groffsky is a fine literary agent, a beautiful person, and a very helpful judge of nonfiction. I am very grateful to both of them for making this book possible.

My appreciation is also extended to IBM and Compaq for making great personal computers and to Multimate Software for producing a marvelous word-processing program that allows a writer to turn out an accelerated flow of words.

Lastly, my greatest appreciation and love to my family for their tolerance and support, even while we all were supposedly vacationing.

Morton I. Hamburg

New York
December 1984

Before You Begin This Book

I have a lawyer friend, Herb, a specialist in corporate litigation, whom I used to run into frequently. Everytime Herb saw me he would tell me how he read everything I wrote about cable television in my *New York Law Journal* monthly columns and compliment me on what a smart fellow I was. He'd always sort of incidentally want to know what telecommunications stock he should buy that week. Each time, out of friendship, I'd give him the name of another telecommunications company whose stock was rising in price, generally exceeding the pace of a rise in the Dow Jones average. I casually told him to buy Viacom International when it was $14 a share, Capital Cities when it sold for $50 a share, and Metromedia when it was "only" $100 a share, and each time I gave him a few reasons why I thought they would increase in market value. The next time I'd see Herb he'd tell me that he "forgot the name of that stock you gave me last time, and do you have any more?" (Viacom is selling these days for about $32 a share, Capital Cities is almost $155 a share, and Metromedia recently went "private" when it was trading for over $300 a share.)

Two years after this series of recommendations, I met him while I was lunching with a client. When I introduced Herb, he told the client that I was a genius—not because I was such an able lawyer, but because I had made him rich by giving him "all those stock tips in the communications

13

industry at just the right time." When I jokingly asked him for a percentage of his profits, he respectfully declined. But when I suggested last year that Herb participate in a private investment I was helping to organize in the broadcasting field, he had a check in my office within a week after receiving the private placement memorandum.

Herb had some substantial assets to begin with (he's a very good lawyer, with some outstanding corporate and personal clients), but he wasn't so untypical when it came to wanting to increase his assets. Many men and women, no matter what field they're in, have a burning desire to learn how to make a lot of money in just a few easy lessons. There are all kinds of books available about, among other financial matters, the buying and selling of real estate, investing in commodities or municipal bonds, and setting up your own venture capital firm. However, I haven't seen anything in the book stores or even much in magazines about how to make a *lot* of money in the most outstanding industry in the world today—telecommunications.

It's true that there are some texts on the technical and constitutional aspects of telecommunications. And one sees daily articles in newspapers and magazines about the breakup of AT&T or the future of cable television, satellites, macro- and microcomputers, cellular mobile radio and cordless telephones, data transmission, long distance services such as MCI and Sprint, and other aspects of telecommunications. Much of this is technical information about how things work, what the latest product is, or what the most recent regulations or act of deregulation may be. However, there is a dearth of nontechnical literature on how the strong-hearted and ambitious can profit from the Information Society we suddenly are living in.

I learned just how little material there is on the subject when I set out to do some research on my own. I found that nobody is out there telling you how the big fellows are doing it, at least not in any comprehensive way. Nobody explains how a Tele-Communications, Inc. (TCI—the largest multiple cable system operator in the United States) was created and operates; how good a company Capital Cities Communications, Inc. is to invest in (in 1984 its common stock was selling on the New York Stock Exchange for up to $170 a share); how a young couple can buy and operate their own radio station; or how Johnny Carson, in addition to having a contract that gives him about $8 million a year (negotiated by his lawyer, Henry "Bombastic" Bushkin), makes even more millions of dollars from a production company funded by his employer, NBC.

There are now over 9000 AM and FM radio stations and over 1100 VHF and UHF television stations in the United States, valued at over $10 billion, but up until now there's been no guide that tells us how to buy and sell broadcasting properties in a creative way, so that any of us may end up with a significant group of stations with a net worth of possibly $20 million *without investing a lot of our own* (or even our spouse's) *money*. It seems time to explain where the best places are to spend one's time, energy, and money to take full advantage of the tremendous growth that telecommunications will enjoy throughout the world in the next decade.

I don't know of any better way today to try to get rich (or at least a little better off) than to become involved in this business in one way or another. I am certain that you can make a great deal of money during the '80s if you buy (or at least buy into) radio and/or television stations, cable systems, or some other telecommunications property in the right way; invest wisely in the right telecommunications company here or abroad; get the right compensation package as a talent or telecommunications executive; or produce a better television program or even a library of videocassettes. The idea is to start *now* and *do it correctly*.

Obviously, there are a lot of companies in the United States and in many other countries of the world presently capitalizing on the growth of telecommunications in various ways. AT&T, IBM, GT&E, ITT, MCI, TCI, Time, Inc., Thorn-EMI, Rolm, Erickssen, NEC, SONY, Philips, Cablevision, and Viacom are only a few examples of the better-known companies involved in the field; I can name hundreds more that are successfully directing their energies and their capital in this industry.

Have you heard of Plessey, Matsushita, or LIN Broadcasting? Or PageAmerica, Scientific-Atlanta, Chyron, or Taft Broadcasting? All of them are publicly held companies involved in the telecommunications industry in one way or another; their securities are traded on the New York or American Stock Exchanges or in the over-the-counter market. All are, in my judgment, extraordinarily good companies to invest in at this time, for reasons which I will explore in this book.

Many of us do not have a lot of money to begin with, but, as I learned and you will see, there are capital and loans out there and available to you that make it possible to get in on the ground floor in the telecommunications business even without putting a lot of money down. I am practically besieged daily by major—and not so major—banks and venture capital firms and by private investors who want to invest or loan up to $1 *billion* in some aspect of the telecommunications business. Many people who are interested in obtaining such financing don't know how to locate the companies that *want* to lend. Moreover, they are not sure what kind of investments or loans they should ask for in order to profit well. This book will tell you!

The problem is not limited only to the uninitiated. Even sophisticated investment managers and investors don't always understand where the golden opportunities are in the telecommunications phenomenon. A lot of people have been coming to me in recent years for more than legal services. They are often primarily interested in getting help in structuring new companies, acquisitions, and investment vehicles in the electronic communications industry. Clearly there is an enormous desire and need for such assistance.

Young people want to know how to get into the Information Society and build a fortune. A lot of people, particularly women, who haven't generally been able to penetrate the broadcasting industry (with some exceptions, such as my wife), want to know from the inside how they can capitalize on their talents in certain aspects of it, especially as performers and managers.

With the greater availability of venture capital and debt financing, a growing number of people want to know how they can start their own communications companies and build up a large net worth. And, of course, *everyone* wants to know the best companies in which to buy stocks, whether through publicly traded shares or private investment (and it doesn't hurt if there are some good tax shelter benefits to such investments).

Naturally, no one today wants to settle for only a few thousand dollars of income or profit. Everyone wants to make *at least* a million, but because of the confusion of technologies in telecommunications, the abundance of regulation, and fear of the establishment, up to now only a few have been very clear about how to do that.

Just look at some of the figures in the cable television portion of the telecommunications industry. There are now nearly 84 million homes in the United States that have at least one television set. They are called "television households." Over 54 million of them are now passed by cable systems. That's 65.1 percent of all households. As of March 1984, there were almost 31 million basic cable television subscribers, or 56.3 percent of the television households passed by cable. There are now more than 18.3 million cable subscribers who also subscribe to at least one pay cable television channel, so that 59.6 percent of basic subscribers also have pay cable. It is forecast by some that by 1990 there will be almost 59 million basic cable subscribers and 43.7 million pay subscribers in the United States alone.

If you multiply those figures by even $10 a month per subscriber, you can see that, notwithstanding some recent adverse publicity about some large urban cable systems and the rapid demise of some cable *programming* networks (not the profitable nonurban cable system *operators*), the cable industry is a pretty healthy one, involving gross revenues of in excess of $2 billion a year in the United States alone. It's a very good business to get into if you know what you can and should be doing to make money in it. This book will show you.

Pick up a copy of *Broadcasting* magazine, the weekly bible of the radio and television business, and read about radio and television stations being bought and sold weekly for up to three and a half times gross income, and sometimes for eight to ten times broadcast operating income, or what is so wonderfully called "cash flow." These sales range from perhaps $200,000 for a radio station in a very small rural area to $325 million for a television station in a market such as New York or Los Angeles. Small groups of six or eight radio stations are being acquired for $40 million, even though organized at a cost of less than half that. Even radio stations in what are known as secondary markets are selling for $6 to $9 million. While the old expression that an FCC license for a broadcasting property is practically a license to steal is not really accurate (although many have stolen without either the FCC or shareholders knowing about it), it *is* true that if you know what you are doing and you are energetic enough, you can do very well by owning a property having such a license. That I can testify to from direct personal experience!

I'm not the genius my friend Herb called me. However, I have been fortunate to have made investments in the telecommunications industry for myself and others that have done well, and I have become active in a communications company that I believe has a great future and is worthy of my time and capital. I feel that my wife and I are on the way to increasing our assets substantially, even if that isn't our main objective in life.

While I can't guarantee that reading this text will make you a telecommunications millionaire, which some have become or are becoming, or even that you will *definitely* increase your net worth at all by getting involved in the telecommunications industry as outlined in this book, I *am* sure that what follows should give the intelligent, dynamic, and determined person enough knowledge to go out and make a *strong* attempt at successfully achieving that goal. Now, let's see what can be done as we share this exploration.

1

The Information Age and What's in It for You

If you use a telephone, you are involved in the telecommunications business. If you are one of the approximately 84 million people in the United States, or hundreds of millions throughout the world, who have at least one television set in their homes and watch some programs daily, and the TV signals are brought to you by commercial or publicly licensed and possibly government-supported television stations (over 1100 in the United States), you are participating in the Information Age. You may be one of the over 30 million subscribers to what is known as basic cable television in this country, and if you have basic cable you probably also subscribe to at least one premium or pay cable channel for which you pay an extra monthly fee or perhaps a special fee on a pay-per-view basis.

In all likelihood, at some point during the day or night you, or at least one member of your family, listen to a radio program carried by at least one of the more than 9000 AM and FM radio stations in the United States—or one of probably double that number in the rest of the world.

If you listen to the radio, use the telephone, watch television, subscribe to basic and pay cable television, exchange data from place to place, or even sail on a ship navigating through one of the oceans, there is a strong probability that some part of such telecommunications is coming to you via satellite.

When you attempt to determine your bank balance, you almost certainly are obtaining such information through a computer linked by telephone lines to one or more other computers containing financial information about you and/or your business. Similarly, if you invest in stocks or bonds of publicly traded corporations, you and your broker utilize communications technology.

You may subscribe to one of the computer-linked information services concerning stocks and bonds traded on the New York, American, Pacific, or other U.S. stock exchanges—or other exchanges around the world. You are able to receive information concerning these stock markets almost instantaneously through a combination of telephone lines, domestic or international communications satellites, fiber optic cables, modems, computers, and terminals, all of which fall within the ambit of exchanging information by means of electronic communications technology.*

Your daily business and personal life probably will require you to obtain information on weather, health, airline schedules, sports results, shopping, dining, or some form of professional activity. You or someone in your business organization may be using some kind of instructional material or attending a seminar for professional or business training through teleconferencing, which uses several forms of communications technology—television, telephone lines, satellites, and computers.

You may simply be one of the growing millions of people who have purchased a videocassette recorder-player, and now you are not only buying or renting videocassettes of some recent movies each week, you are also transferring the time when you view certain television programs each day. For business purposes, including training of personnel, you may be utilizing laser-operated videodisc equip-

---

*The glossary at the back of this book contains brief explanations of various telecommunications terms. Don't hesitate to use it when you feel the need.

ment. Although you probably don't participate yet, the time may rapidly be approaching when you will be video-recording pictures and sound for displaying your products or real estate that you are selling, keeping a record of your children's growth, improving your speechmaking capabilities, or simply for showing friends on the other side of the world how you and your family look these days.

Soon an enormous number of people in the United States and throughout the world will be listening to and viewing such material, as well as regular television programs broadcast over the air in stereophonic sound (allowing dual language programming) and with high-density, absolutely clear and colorful pictures. This is probably the most significant economic advance in day-to-day communications for entertainment purposes since the first sale of regular television sets in the 1950s.

You may also be one of the number of people, professional and otherwise, who feel a growing need to have a cellular mobile telephone in your car. Perhaps you do a great deal of moving around, locally as well as to other cities, and want to make sure that your patients, customers, employees, or simply your wife or husband can page you at any time. If so, you are participating in, or are about to participate in, another aspect of the expanding telecommunications business, paging devices.

Or perhaps you aspire to become one of the millions of people professionally involved in the telecommunications industry. You may be talented in either performing or producing some kind of programming used in the telecommunications business. This clearly is communication in its truest form and makes you an active participant in the Information Age.

Never in the world's history have so many individuals and organizations been able to interact on such a vast scale. By the year 2000, electronic information technology will have made vast changes in American business, manufacturing, education, family, political, and home life. I'm sure that the same can be said for most of the rest of the world.

It is all very expensive technology, and the cost of researching and developing it is astronomical. So is the cost of building plant and equipment. However, an investment in such technology now may be vastly rewarding in the future.

What's in It for You?

If there is so much money being spent in the telecommunications industry, then it stands to reason that there is a great deal of money being earned by a lot of corporations and individuals in the business. Vast fortunes will be made by many if what is being said about the future comes true. We are no longer talking of *millions* of dollars or even of *billions*. When projections are made of revenues for companies involved in the world-wide dissemination of information through voice and data by telephones alone, we are now discussing the future in terms of *trillions* of dollars, or the fiscal equivalent thereof in other than U.S. currency.

What's in it for you? Probably *a lot*—if you really want it. If you simply want to invest some part of your assets, small or large, in public corporations, limited partnerships, or privately held entities in the telecommunications industry, with careful investment—buying and selling—you should be able to increase those assets, perhaps to the point of making a million dollars or more. It doesn't necessarily take a lot of money to make a lot more if you invest wisely and well in this constantly expanding industry.

If you want to become directly and actively involved in a telecommunications entity, there are still vast opportunities for you to do so. You can still buy radio or television stations in the United States, or even whole groups of stations, perhaps not alone, but with the help of other investors and banks. You can buy and operate cable television systems, if not the large ones, then possibly some "classic" systems that throw off good cash flow and have potential capital gains value. You may even want to set up your own satellite programming network.

There still is room for getting into the cellular telephone business and for selling paging devices and services. Now that AT&T has been broken up, there are great opportunities for marketing, installing, and servicing telephones and associated equipment. You may even have ideas about how communications technology that no one else has yet thought of can be utilized, and all you need is some direction about how to put your ideas into a profitable organization. You will find it here.

If you own or are in a bank or venture capital or investment banking firm, there are tremendous opportunities for you to get together with people already in or planning to enter the telecommunications field. You can attain a higher position or increase your own asset value by arranging for loans to, or privately investing in, some of the thousands of telecommunications companies springing up like wildfire throughout the world.

The enormous need for programming of all sorts in the many areas of telecommunications in this country and abroad, now and for many years to come, may make you want to get into the business as a talent in front of or behind the cameras and microphones, acting, moderating, newscasting, directing, or producing.

If you have management capabilities (or can get training in one of the business schools), the need for your talents is and will be vast. The rewards for executives, male and female, can make you as well-off as many of the high-priced men and women already in the field. It is a field in which there is currently a *minimum availability* for a *maximum need*.

You may be attending a law school and considering a career in communications or entertainment law or in industry rather than in a law firm. The constantly expanding telecommunications industry requires not only good lawyers, but executives and business affairs directors with legal backgrounds, brokers who can assist in creating good acquisitions, and talent agents who know what contracts and taxes are all about.

Financial and accounting acumen is vital to every industry, but people with such abilities who are inclined toward telecommunications should reap significant rewards by getting into the field. This can be done either by going to work for one of the corporations in the telecommunications business or by assisting clients involved in the field through a professional practice such as accounting or the law.

There are limitless opportunities for the initiated and uninitiated in this Information Age to make a good living, even to get rich.

One of the nicest things about the industry—certainly for my wife and daughter and possibly for you, too—is that there are vast opportunities for women in every phase. The demand for talent is so great that the prejudices against females that have long endured in many areas of the industry are now breaking down, and I truly believe that the way to success in telecommunications is no longer restricted by sex. As we shall see, women stand out on every level, but there may be some better ways than others for them to achieve in the field.

Making Money in Telecommunications—Some Basic Questions

If so much money is being and will be made by so many in the telecommunications business, you must want to get your own piece of the pie. Well, I think I can help you do that, whether you want to be a passive investor or play an active role in the industry. You can do it either way (or both ways), but first you have to do a little self-examination.

If your objective is to make money and/or have a career in telecommunications, there are a number of ways in which you can do it. First, however, as I tell all of my clients (even my wife, although she won't ever really let me consider her as a client), you have to make some evaluations on your own. Basically, you must ask yourself the following questions:

• Do I have the time to devote to playing an active role in an organization, new or existing, involved in telecommunications? Or should I simply be an investor in entities presently in or about to enter the telecommunications business?

• If I have or really want to make the time to devote to an active role, do I have the *capacity* (i.e., talent, training) to get into some aspect of the industry or to draw in others who have the ability to do so and learn from them?

• Do I have the necessary *capital*, or can I attract funds to make the investment, either passively or actively?

• What aspect of the telecommunications industry most *interests* me?

• Perhaps most of all, do I have the *desire, patience, and fortitude* to see the matter through, again either from a passive investment or active participation point of view?

If you have the *capacity, time, financial clout, interest, qualifications, patience, fortitude,* and *desire* to become actively involved in some aspect of telecommunications, then you had better get started *right away,* because the decade is passing rapidly and a lot of people are already there. If your self-evaluation comes up negative on the active side, then recognize how important it is for you at least to become an investor in some aspect of the industry through purchasing securities in at least one company—but hopefully more—involved in telecommunications, and to do so right away before the old adage "Buy low, sell high" becomes impossible in this field.

Regardless of the result of your self-evaluation, in order to make any investment of your time, talent, and/or money in a telecommunications entity worthwhile, you must have some idea of what has happened, what is happening now, and what probably will happen in the future in the industry. This book will tell you.

Clearly there is a passage of time between the writing and the publishing of any book, and some of the things that I say here may be affected by intervening events that are not foreseeable. I obviously give no guarantees about anything recommended in this text, but I can tell you that I firmly believe that my thoughts on how profits will be made in the telecommunications industry are valid and that I do put my (or my clients') money where my mouth is, as they say.

If you do not think you are interested in or don't, in your estimation, have enough capital to make a meaningful investment in public companies, I suggest that you read through the next chapter anyway. As you learn about some of the companies that are or will be doing well in various aspects of the telecommunications business, you may find you can and want to become actively involved. You can examine one or more of those companies in greater depth to learn how they are doing it. Then maybe you can do the same, perhaps on a much smaller scale. You may want to get a job in one of those companies so that you can learn about the business. (Appendix B at the end of this book lists the top executives and personnel managers at most of the companies—at the time this was written—so that you can contact one or more of them.)

Now let's go on to look at some of the telecommunications companies that you might want to review as investments.

2

The 30 Best Telecommunications Securities to Buy—and Why

The headlines scream out from newspapers, magazines, stock reports, and the financial press seemingly every day:

"RCA Has Record 2nd-Quarter Net"
"Motorola Net Surges 92.2%"
"Ameritech, NYNEX Up"
"AT&T Earnings Jump; Assure 30¢ Dividend"
"Cable Programming Grows as Force in TV
    Syndication"

All the signals are there about how the right investment in the telecommunications industry pays off, how the securities of so many corporations involved in the field appear to be undervalued, and how many companies are headed for success in the years to come. Why, then, isn't there a mad surge to buy these securities?

Well, there are a number of companies whose securities are appreciated by knowledgeable investors, but even some sophisticated fund managers are so unsure about such investments that they pass by many of the good telecommunications stocks to invest vast sums in securities of companies that couldn't possibly perform as they expect them to.

Such uncertainty and ignorance, however, open an even greater opportunity to those who get into some of these

28

telecommunications stocks *now*. Their growth is just beginning, and assuming that nothing catastrophic occurs in the world's economy between now and 1990, we will see in telecommunications the largest growth of any industry in the United States, and probably throughout the world. Such growth has to be registered not only in terms of physical plant and use of the technology, but also in terms of revenues and earnings and consequent increases in the prices of publicly traded securities.

In order to create a portfolio of successful securities, you must, of course, have some capital to begin with. The mechanics of trading in telecommunications securities I'll leave to those whose profession it is to advise customers in doing so. But you don't need vast sums to begin with to build up what anyone would regard as substantial net worth if you invest wisely and well. I'm sure that no one can tell you in *any* text what to do about particular investments from day to day over a period of years, but you can follow some general principles and even explore specific possibilities in order to meet your long-term objectives.

The bottom line to choosing the best companies to invest in over a period of time is to keep a close watch on their progress by regularly following the market prices, by examining all the material you can obtain from the privately held companies, and by reviewing the annual reports of publicly held companies—and even attending their annual shareholders' meetings. Most of all, you must have, as the late mayor of New York, Fiorello LaGuardia, used to say, *patience* and *fortitude*!

I can't tell you about every telecommunications company that you might invest in to increase your assets by 1990, but I don't mind naming and examining a few that I think are the best to buy at this time. Since *best* is measured purely by my own criteria, you will be interested in those criteria. I'll tell you about them, but first let's see which are the thirty best buys.

The 30 Best Publicly Held Telecommunications Companies to Buy Into

If you want to make some money between now and 1990, I'd suggest that you buy stock in at least one, if not more, of the following telecommunications companies. The abbreviations in parentheses after each name indicate whether the company is listed on the New York Stock Exchange, the American Stock Exchange, or traded on the over-the-counter market.

1. Time, Inc. (NYSE)
2. Capital Cities Communications, Inc. (NYSE)
3. Zenith Electronics Corporation (NYSE)
4. RCA Corporation (NYSE)
5. Motorola, Inc. (NYSE)
6. Matsushita Electric Industrial Company, Ltd. (NYSE)
7. The Plessey Company plc. (NYSE)
8. A. H. Belo Corporation (NYSE)
9. Advanced Systems, Inc. (NYSE)
10. Chyron Corporation (OTC)
11. LIN Broadcasting Corporation (OTC)
12. Taft Broadcasting Corporation (NYSE)
13. Cox Communications, Inc. (NYSE)
14. Malrite Communications Group, Inc. (OTC)
15. Viacom International, Inc. (NYSE)
16. Lorimar Corporation (ASE)
17. PageAmerica Group, Inc. (OTC)
18. Tele-Communications, Inc. (OTC)
19. Rogers Cablesystems, Inc. (ASE)
20. American Broadcasting Companies, Inc. (NYSE)
21. CBS, Inc. (NYSE)
22. Telepictures Corporation (OTC)
23. Scientific-Atlanta, Inc. (NYSE)
24. GTE Corporation (NYSE)
25. Tribune Company (NYSE)

26. NYNEX Corporation (NYSE)
27. MCI Communications Corporation (OTC)
28. American Telephone and Telegraph Corporation (NYSE)
29. International Business Machines Corporation (NYSE)
30. Dow Jones and Company (NYSE)

There are literally hundreds of corporations throughout the world today engaged in some aspect of the business of telecommunications. There will be hundreds more created before the decade is over. For the purpose of our exploration, however, I believe these thirty publicly held telecommunications companies offer the best securities to purchase now to hold for long-term gains. Naturally, events may occur in and out of these companies, in the world economy, or in the stock market generally that will cause any one or all of them to show little or no growth during the time period that I've arbitrarily set as our objective for making a significant profit—the year 1990.

I've listed the companies in the order of my preference. I haven't indicated what the current market price of each security is because this isn't a weekly newsletter and prices may change significantly between my stating these recommendations and your being able to read this. I hope that you will consult with your own investment adviser or broker (if you aren't one yourself) to learn of any new events that have occurred involving these companies that might affect their value.

You won't find among this group any stocks that sell for $1 or less in the over-the-counter market, the so-called "penny stocks." In fact, many of these securities have been selling for from $60 to $80 a share, and some for much more. I can't (and won't) tell you now to take $1000, buy 10,000 shares of X Communications Corporation, and make $1 million in a year when the stock goes up to $100 a share. I haven't seen any companies that I think will do that, and

I'm too conservative a guy to believe that anyone else has found such an entity.

However, I believe that you can buy a certain number of shares in companies with continuing growth and turn those investments into significantly profitable assets. Yes, you can even do it with a $160 stock, if somewhere along the way there is a splitting of the shares of such a corporation and you wind up with 400 shares worth $40 or $50 a share instead of 100 at $160 or $200 a share.

There are some lower priced stocks in this list. Some of them have had their ups and downs, and hopefully, when and if you call your broker to inquire about buying, one or more will be low enough to be bargains. If you have a comprehensive portfolio of these stocks, you can do enough trading with $1000 capital as an initial investment (assuming that you buy some of the lower priced ones, or some that you suspect will split their shares in the near future, and keep selling and buying as you reinvest your profits) or $50,000 (which you might want to use to purchase only one or two outstanding stocks that you think are substantially going to increase in value over a period of time) to make a meaningful entry into the telecommunications market. I can tell you that you are not going to get rich quick. Hopefully, you will make some money, and in the long run you *might* get rich.

I haven't given long-winded explanations of why I believe these companies are the best. If they interest you, you and your financial adviser can explore them in detail. Much of my information is derived from 1984 reports and discussions. It is pertinent to *my* exploration of the best companies, but it is not as up-to-date as it would be if you were to read this material today. You can obtain more information about any of the corporations from Standard and Poor's corporation sheets, from other good financial news sources at your favorite brokerage house, or even at a public library. In certain cases, my inclusion of a company as a "best" may be just a visceral feeling that I have for some of the technology that company is developing or because I know the top management and suspect that they will keep building up their company's profits. By no means do I invest my money, my clients' money, or any trust funds on a mere whim, and you shouldn't either.

The Criteria for Anyone Choosing the Best Companies

I'll tell you what I do to choose what I think are the best telecommunications securities to buy; you might want to try to think along the same lines. I followed four criteria in evaluating every company listed here. You may not necessarily want to duplicate these guidelines, but you should give serious thought to them.

1. THE GAME PLAN

First and foremost among my criteria is what kind of *game* or *business plan* does the corporation profess or appear to have for the long-range future? Is it indulging in the latest vogue simply to get into something that other companies with bigger and better resources are already very much involved in? Has the company the knowledge, strength, and willingness to see its plan through to large rewards not only for its leading executives, but for the holders of the company's securities as well?

In order to know whether or not a company has a valid business or game plan, you must have some idea of what has gone on in the areas that it may be plunging into, as well as have an awareness of the basic soundness of such a plunge. Learn what you can about the industry as well as the individual company before purchasing any security. For instance, if in 1980 you had asked some knowledgeable people in the field, or had learned more about the industry, you could have seen that the RCA Corporation was making a big mistake in significantly committing time, money, and people to manufacturing and marketing a videodisc player that used a very fine magnetic needle rather than a laser beam to play videodiscs. At that time, some people (including me) were predicting that RCA would end up writing off at least a $90 million loss when it gave up the operation, as it appeared it surely would.

Herb Schlosser, the former head of NBC and then the man in charge of RCA's videodisc division, was telling many people about the bright future for their operation. He got caught up in the programming possibilities of videodisc technology. I had been involved almost twenty years earlier in representing a public company with the wonderful name Responsive Environments Corporation, which was in the computer-assisted learning systems industry in the United States and England. I knew of the difficulties experienced then by many large and small companies, not only in marketing such technology, but in developing the kind of programming (now called "software") required to sell the machinery (or "hardware," as it is now called).

Almost anyone could have seen that significant advances had been made in the development of videotape and videocassette recorders-players. Very few people in this country, or for that matter anywhere else in the world, want to buy a machine that plays only an established program (as opposed to one that they can record themselves), and that can be played only as the magnetic needle progresses along the disc without any interactive capability. Videodiscs played by laser beam permit you to go back and forth along the disc to any one of thousands of locations, giving you random access and long life for the disc. With a videocassette recorder a prerecorded tape can be played, other material can be recorded off the air, you can use a camera as you choose.

When RCA's game plan was publicly revealed, many failed to see that it wasn't a good one. RCA recently announced the closing of their videodisc operation, with an apparent loss of about $500 million. However, as we shall see, RCA has learned a few lessons, and I think they have a much better business plan now. They're still a pretty good company.

It's not easy to determine the business plans of public companies, but I think you can. Get hold of all the recent annual reports of the company you're interested in, any promotional materials they put out, the quarterly statement, and any other literature that you can find about them. Read business magazines such as *Forbes, Barron's,* and *Fortune*; newspapers like the *Wall Street Journal* and the *New York Times* for any stories on the companies or their area of the industry. Attend shareholders' meetings whenever you can. Once you make your determination of what a company's plan is, then try to evaluate its worthiness. Hopefully, with some knowledge and insight, you can do so yourself, or you may want to consult some experts who can.

2. NO "HYPE"

Clearly, knowledge of what has happened in certain areas of telecommunications gives you a better chance of evaluating what a company may be getting into. It also provides a basis for determining whether a good company may be led into a business that it can't possibly hope to succeed in by pure public relations, or "hype," as some like to call it. CBS, Inc. has in recent years fallen victim to this syndrome on at least two separate occasions. Some years ago, CBS, a most successful broadcasting station licensee, program producer, television and radio network operator, record producer and distributor, and publisher, decided to get into some new technology. It settled on a system called EVR, which not only didn't work well as a form of videodisc player—it just didn't work! Again, millions of dollars went down the drain, and a lot of it was spent on advertising and promotion of a nonworking piece of equipment.

CBS, Inc. was forced by the government a number of

years ago to spin off a company through which it owned certain cable systems and through which it was also syndicating television programs to TV stations. The spin-off company evolved into the presently successful Viacom International, Inc., which we'll discuss later. When cable television came into vogue, CBS started a division called CBS Cable to disseminate cultural programming to cable television systems, apparently without considering that few cable systems had been built in the large urban areas, where culture fans are most numerous. CBS Cable failed in spite of some wonderful and expensive programs, graphics of superb quality, great advertising, and a number of marvelous public relations parties thrown by its well-meaning executives. CBS Cable failed simply because it came before its time. If it had a business plan, it wasn't based on practical theories and facts, only upon hope and hype. Anyone who knew anything about the cable industry at that time could readily see that. It's been estimated that CBS, Inc. wrote off about $35 million on that one.

Try to separate out fact from hope and to see through hype. Don't invest on some vague hope unless you or your adviser are qualified to evaluate, and try to get an opinion as to whether or not the projected business can really succeed.

3. AVAILABILITY OF CAPITAL

The third most important feature I look for in deciding on investment in securities in the telecommunications industry is the *availability of capital* to the company being considered. In order to get into new areas of telecommunications, in many cases great resources must be committed by corporations for research and development, for building and buying plant and equipment, and for attracting talented personnel. This applies to companies just getting into the field as well as to many entities that have had vast experience in the area. It is as true of large corporations as it is of small ones. The staying power required is awesome, but, of course, the rewards can be great, and some investment bankers, banks, and venture capitalists have in recent years made tremendous amounts of money available in one way or another to such companies.

The potential investor must have some idea of what kind of capital the company presently has or will have during the term of its intended development to carry out its plans, and of what will be made available to that company for the future. This makes it important to determine not only what kind of promises of capital have been made, but *who* has made them. It is very important to know what investment banking firm, what bank, which venture capital organization, or even what investing individual is associated with the company.

In order to evaluate this, you must look at the company's balance sheet and its most recent annual report, even its filings with the SEC, if you can obtain all of these. See how much working capital the company has and what kind of capitalization it has established so that you can determine if it will be able to go back to its banks for the money to do what it says it wants to do, if it can safely float a new bond or debenture issue to obtain the funds, or if it must issue some more equity to dilute the position of its common or preferred stockholders. You may even be able to obtain information from company executives or from analysts of the company who have talked to some of its people.

4. THE KEY PERSONNEL

Last, and perhaps most importantly for me, is the matter of who the *key people* are in the corporation or other organization in which you may invest. How many times have you seen seemingly good companies with vast resources wind up attempting to carry out a perfectly good business plan with the wrong personnel? On the other hand, how many times have you seen strong, intelligent, well-organized, and dynamic people move companies or partnerships into areas of great achievement when there seemed to be only inertia and misdirection before? Nowhere have I seen better examples of the latter than in the communications business. While no one person can make a company these days, I do think that in such a dynamic industry, forceful men and women can mean the difference between the success or failure of a company, or at least a division of one.

So I like to find out as much as I can about the leadership of a company, including where someone has been, what his or her thinking is about areas the company is in or is contemplating going into, maybe even what kind of failures have been sustained (and overcome), and whether or not such a person is being honest in the projections being made or is simply trying to "hype" his company to success. I've actually been lucky enough to have lunch with Terry Elkes, the president of Viacom, in his office, to discuss not only his outlook on the future of his company, but on the entire telecommunications industry. When I first became interested in Capital Cities Communications, I made it a point to meet Tom Murphy, the chairman, at a cable television convention, and to discuss with him what he saw as the future of cable television within his company shortly after its acquisition of Cablecom General's cable television systems.

I try to read every story I can in the leading business magazines about the top executives in the telecommunications companies, and whenever and wherever I can, I try to meet them and find out what they're like and what they really think about the future of their companies. I don't think it is a securities law violation to meet with the chairman of a public company to find out whether he is talking about his company's products for public relations purposes or if he really believes in them and his own management (and his business plan) and then go out and purchase some of his company's stock. I actually did that with two telecommunications companies—and did pretty well, too.

I realize that making these determinations about public companies isn't easy for just anyone to try, whether you're living in New York or Los Angeles or in Green Bay, Wisconsin, or Santa Fe, New Mexico. You have to make an effort yourself, and find the persons and materials that can be helpful to you in your investment planning, and you must do it when you feel the moment is right, which usually is immediately.

Why I Like These 30 Telecommunications Companies—and You Should, Too

In the pages that follow I will tell you why I like the thirty telecommunications companies I have listed here and how I've applied my criteria to them to make them my best buys. You may not agree with every one, and you shouldn't just take my word for them, anyway. You (and your stock broker or financial adviser) should have some ideas of your own in choosing what securities to buy in telecommunications, or in any industry. But you'll see why I say that these seem the best to me.

TIME, INC. AND CAPCITIES

My two favorite public companies involved in the communications industry to invest in today are Time, Inc. (NYSE) and Capital Cities Communications, Inc. (NYSE). These two companies are very similar in their outlook, but they are quite dissimilar in their operations. Both, however, have the same possibilities for continued capital growth, and for some of the same reasons, the most essential of which is that *they are both heavily involved with the cable television industry.*

In order to appreciate the value of these companies, you

must understand a little bit about the cable television industry and also about pay television. It's now some thirty years since the cable television industry was started in the United States, initially in Pennsylvania by some television set retailers who could not bring in television signals from television stations in the larger cities except by constructing large master antennae on the tops of mountains and then running cable to their stores and on to customers' homes. They then made the people who bought the sets and had the cables installed in their homes subscribers to their cable television systems.

Presently there are some 54.5 million homes passed in the United States by cable television systems, or 65.1 percent of all households in which there is at least one television set. There are approximately 30.7 million subscribers to basic cable television, or only 36.7 percent of television households with basic cable. About twelve years ago, a company called Home Box Office, Inc. introduced the modern premium pay cable era. By paying an extra fee, subscribers to basic cable television systems were able to obtain certain programming (mostly motion pictures) on a monthly basis without any interruptions and without advertising.

Today there are some 18.3 million subscribers to pay cable, but they constitute only 21.8 percent of households with television sets and 33.6 percent of homes passed by cable systems. The large trunk cables that carry the various cable television channels from the cable company's central location, usually called the head end, are laid in streets on which there are so many homes and/or apartment houses and thus "pass by" so many dwellings with potential subscribers to the cable system. All the cable company then has to do is connect the individual home or apartment house to the trunk line, install the smaller cables in the dwelling, and permit the subscriber to tune into the cable system, usually by means of a cable converter attached to one or more television sets.

Almost 60 percent of subscribers to basic cable service also subscribe to pay cable. It's now estimated that by 1990 there will be at least 50 million basic subscribers in the United States, and about 33 million of these will subscribe to at least one pay cable service.

I believe that U.S. cable television system owners will build all of the systems that are presently franchised or will be franchised in the next few years. I also believe that the majority of our population will want to subscribe to cable in their homes (and probably their offices) and that they probably also will subscribe to at least one pay cable service. You also have to believe that similar growth will take place in developed countries abroad.

While I've never believed that people needed or would even want 102 channels to come into their homes, hotels, offices, or business establishments, they will want (and need) a significant number of channels. They will want them not only for various kinds of entertainment as presently provided, or for sports and news networks, but for police, fire, and medical alerts, for certain kinds of data and specialized information, and for shopping, banking, and other services. They will also want to assuage their thirst for knowledge, so educational materials (as broadly interpreted) will be delivered and interacted upon via cable, probably using fiber optics.

As I have said, you have to believe in the future of cable television in order to agree with some of my recommendations for investment, especially when it comes to the two leading companies. It is also true for some of the other possible investments, and I'll try to point out to you how this occurs as we go along.

Time, Inc. was founded as a publishing company by Henry Luce in 1922 to publish *Time,* "the weekly newsmagazine," and it has done that quite successfully for over sixty years. It has now become the largest publisher of "general circulation" magazines in the United States. (*Sports Illustrated, People, Money, Life,* and *Fortune* are some of its other magazines.) Most significantly for our purposes, in November 1978 it purchased American Television and Communications Corporation, which now operates cable television systems in 467 franchised areas in thirty-one states with over 2.25 million subscribers.

Time started Home Box Office in 1972, and now HBO is in 13.5 million households. Many New Yorkers who subscribe to cable television through Manhattan Cable Television (which has the franchise granted by New York City's Board of Estimate for the lower half of Manhattan) and who also subscribe to HBO are surprised to learn that all the companies involved are owned by Time.

For a while, Time, Inc. looked as if it were going to be simply a print publishing and forest products company, but because it very wisely took advantage of the telecommunications explosion, its emphasis in recent years has been on its large income-producing Video Group. In January 1984, Time fortunately spun off its forest products business, which it had entered through the acquisition of Temple-Eastex in 1978.

Time has made singular strides upward under the guidance of J. Richard Munro, its current president, and Gerald M. Levin, a forty-four-year-old former lawyer who, as head of HBO, made the momentous decision to deliver pay cable programming to cable systems throughout the United States by using domestic communications satellites. Levin is now executive vice-president of Time and very much concerned with the Video Group, the essential part of the corporation. The Video Group accounted for almost one-half of Time's revenues in 1982 and 1983 and apparently produced more income from continuing operations than all the other Time activities. Levin is also involved in the long-range planning for the entire corporation.

Home Box Office has had a succession of bright young men involved in its development. Its growth was somewhat stunted in 1984, primarily because of the slowed growth in the building of basic cable systems, particularly in the larger urban areas of the United States. The Video Group is now being supervised by N. J. Nicholas, a very good financial executive who once managed Manhattan Cable Television in New York City. HBO's current chairman and chief executive officer, Michael Fuchs, is an aggressive, knowledgeable, highly intelligent man (a former lawyer and business affairs director) who seems to know where his operation has to go, near and long term, and who has the resources to carry out the division's own business plan. HBO has gone a little wild in recent days trying to make its own movies and committing itself to very expensive deals to get movie product from some of the large film companies like Columbia Pictures. HBO is currently one of three partners with CBS, Inc. and Columbia Pictures Industries, Inc. in Tri-Star Pictures. Frank Biondi, who made some of these movie deals as chairman of HBO, recently stepped down, presumably because of differences of opinion with some senior Time, Inc. executives, and has gone to a film company, Columbia.

Notwithstanding its current lack of growth as compared to the previous few years, HBO should maintain its wide leadership in the pay cable field and contribute appreciable dollars to Time, Inc.'s revenues and earnings during the rest of this decade. It has plenty of competition from other pay cable networks and videocassette rentals, but when the cable industry truly reaches its capacity, HBO will have a great lead for pay television, including the pay-per-view sector.

In 1983, Time had revenues of $2.717 billion, down from $3.564 billion in 1982 and $3.296 billion in 1981. However, its nine months 1984 revenues exceeded $2.2 billion, up from $1.9 billion in the corresponding 1983 period (a 13 percent rise). Total 1984 earnings are projected at $3.25 a share, a significant increase from the $2.25 per share earned in 1983.

For some strange reason, Time, Inc.'s common stock has been trading in the $40 range, and I really can't understand why it remains so "reasonable." I fully expect this to be one of the leading telecommunications companies of the decade, and if it doesn't make any rash mistakes, it will be. For long-term capital growth, there may not be a better investment in the industry.

My other favorite company for investment in the field is Capital Cities Communications, Inc. It seems odd to say that a company whose common stock is selling for around $160 a share is a good investment for growth, but it is true. I've been buying it for a long time; I never hesitated to suggest that its stock be bought when it was $60 a share, or $90, or even $120. Between 1973 and 1983, Capital Cities had a ten-year compounded growth rate of 20 percent, and it's become a significant and growing factor not only in radio and television station ownership, but in publishing and in cable television system operation. CCCI now owns six television stations, four of which rank number one in their markets (in Houston, Philadelphia, Buffalo, and Fresno). CCCI recently agreed to purchase WFTS-TV, Channel 28, in Tampa–St. Petersburg, Florida, the seventeenth largest market in the United States, for $30 million. The company also presently owns six AM and six FM radio stations (a company with which I am involved purchased an AM-FM combination from CCCI in December 1983, in Albany, New York, the AM being the first radio station CCCI acquired in 1954), and they currently are looking to purchase two more radio stations in larger markets. Now that the FCC has increased the number of stations that any company may own, I am sure that CCCI will become one of the largest group owners in the industry.

In 1983 Capital Cities' broadcasting revenues increased by 7 percent to $235.8 million, with operating income of $122.6 million. In 1984, the revenues of the Broadcasting Division were up significantly, and earnings were up also.

CCCI has operated in an intelligent, somewhat conservative, and highly lucrative fashion under excellent leadership for an appreciable period of time. It also has in my view made one of the best acquisitions in the field of telecommunications, a coup which it has not strongly publicized but which will cause it to reap even greater rewards by the end of this decade.

There's a little story that goes along with my outlook on Capital Cities. In 1979 I was asked by a brilliant and most successful young man who had made a great deal of money on Wall Street and was then acting as an executive committee chairman of a large motion picture company to consult for his company to help get it into the cable television business and, incidentally, to assist in setting up a videocassette division. In addition to reviewing much of the company's operations in California, Illinois, and New York, I spent a good deal of time carefully examining various cable television companies, even talking to people in the industry about possibly joining forces with the larger company.

Thereafter I recommended that the company make an outright purchase of either of two companies in the cable field, one of which was Cablecom-General, Inc., a subsidiary of RKO General, Inc., itself a subsidiary of the General Tire and Rubber Company (now GenCorp). I put into a memorandum my reasons for suggesting the purchase of the cable company—all of Cablecom-General's cable systems were in nonurban areas, it didn't have dynamic management, it wasn't involved in seeking franchises in any major cities so would not have to expend a great deal of capital to try to obtain any new franchises and, even more importantly, to build them if successful. I said that I thought the company needed some good management at the top and some marketing expertise. I felt its operations produced enough cash flow then to warrant paying at least $500 per subscriber for the company, or $100 million, which I thought would be well worth it in view of the fact that

Cablecom-General's operations would be valued at over $200 million in the not-too-distant future because of its operations, its potential, and what was going on in the cable industry.

My clients told me they had the feeling that I was right about RKO wanting to sell Cablecom-General and that I was probably right about the $100 million purchase price, but they thought such a price was too much for their company to pay. It was suggested that they would do better if they got into the cable business by starting a new pay cable satellite network programming company together with some other large motion picture companies. I told them that while I wasn't an antitrust expert and wasn't giving a legal opinion, I thought they were wrong in passing up Cablecom-General, and if they tried to compete with Home Box Office on such a large picture company "conspiracy" basis, the Justice Department and HBO would probably have it enjoined on antitrust grounds.

A few months later, Capital Cities Communications, Inc., after a careful study of the cable industry and particularly of Cablecom-General, purchased Cablecom-General's cable operations for $139 million. As of December 31, 1983, the Cable Television Division of CCCI had interests in fifty-three cable television systems serving 349,520 basic subscribers with 249,850 premium units. The systems are in sixteen different states, none in any major urban centers. All are built systems, and all are producing cash flow. In 1983, revenues from the cable operations totaled $67 million, a 26 percent increase over 1982. All of their acquired systems produced $19.4 million of cash flow. The division's new systems produced $2.3 million of operating cash flow for the company. Total operating income went from $1.7 million to $2.1 million from 1982 to 1983.

Using an industry standard of possibly paying $1000 per subscriber for a cable system today, or even ten times operating cash flow, it's clear that CCCI made a helluva buy; my prediction of valuation of at least $200 million has come true. Under the very able leadership of William R. James, the president of CCCI Cable Television Division, the cable operations have made very conservative but considerable progress, and they are almost certain to continue in their successful direction. I doubt that CCCI would sell its cable operations today for anything less than $250 million; the long-term benefit to the company would seem to support its $160 common stock price, even much more.

Speaking of outstanding management, Capital Cities is led by Thomas S. Murphy, its board chairman, and David B. Burke, its president, who operate their company in a very economical manner. Both are prime examples of dynamic managers with a great understanding of the communications business and outstanding foresight.

By the way, after I had ceased consulting with them, that motion picture company did join with three other large film companies and tried to start a pay cable network called "Premiere." On December 31, 1983, the United States District Court for the Southern District of New York, at the request of the U.S. Department of Justice, enjoined the network from continuing its operations on antitrust grounds.

It's clear to me that as the broadcasting and cable television industries grow in this country, and as the publishing business keeps producing significant earnings for those who meet the diverse needs and interests of a constantly growing reading public, companies with intelligent game plans, capital, and dynamic management, such as Time, Inc. and Capital Cities Communications, Inc. will see the value of their securities increase many times by the end of this decade.

ZENITH AND RCA

When you think of telecommunications companies, you have to consider not only those that sell services, but also those that manufacture and market the technology. In certain cases, companies may have divisions that cross over into both areas. Zenith Electronics Corporation (NYSE) and RCA Corporation (NYSE) are two examples of such companies, although their divisions are quite different except for one essential—the manufacturing of television sets.

Zenith is primarily engaged in the design, development, manufacture, and sale of consumer electronics and related products, and their main products these days are color television sets. They also make and market videocassette recorders, color video cameras, video components, and decoders. Zenith also makes addressable pay cable television decoders, electronic studio equipment, and other technology used in broadcasting and cablecasting. In 1983 it had net sales of $1.361 billion and net income before income taxes of $87.5 million, going from a loss per share in 1982 of $1.28 to fully diluted net income per share of $2.11 in 1983. Through September 30, 1984, Zenith had net sales of $1.2 billion, up almost $200 million from the corresponding period in 1983. Year-end figures should be even greater.

RCA Corporation has had a lot of trouble over the past decade. It's gone into and out of a number of ventures (including the videodisc) without success and at considerable expense. Its competition from the Japanese and others in electronics has been awesome. It has had difficulties with its top management. Notwithstanding all of this, RCA has kept itself among the leading manufacturing companies in the telecommunications industry, and it still holds on as the parent corporation for one of the three leading television networks in the United States, the National Broadcasting Company.

RCA sales in 1983 rose to a record $8.98 billion. Net income increased to $240.8 million, or $2.10 per common share. Similar growth occurred in 1984. Its satellites, earth stations, and broadcast systems are prevalent in the United States and in many other parts of the world. It has an outstanding board of directors and now, under the leadership of Thornton Bradshaw as chairman and Robert R. Frederick as president and chief operating officer, its management is ready to carry out what appears to be a forward-looking business plan. Its stock is selling at a rather low price-earnings ratio at this time.

The main reason these are two of the best companies involved in the manufacturing of telecommunications technology for investment, however, is common to both. In a recent development, the Federal Communications Commission favored a system developed by Zenith for the transmission of television programming with stereo sound. Earlier that system had been unanimously endorsed by an industry subcommittee. The system not only permits programs to be heard in stereo, thus enhancing the value of many kinds of television programming, but it also allows programs to be telecast in at least two languages simultaneously, thus opening up viewing of many American-produced programs to a growing population here for whom English is a second language. It will also enhance the marketability of programs produced in other countries.

Stereo television will be the single largest producer of communications products sales in the world during the next six years. Just as practically no home in the United States is without at least óne television set today and many have multiple sets, at least one of which is color, so every home and many places of business and entertainment will have at least one color set with built-in stereo sound. This means that a whole new market is being created with many billions of dollars of sales and earnings to be made. From my vantage point, the two U.S. companies best prepared to take advantage of this market are Zenith Electronics Corporation and RCA Corporation. The value of their shares will multiply with such sales, and those who buy into these corporations now may share *pro rata* with the management and other shareholders in such growth.

MOTOROLA

When I was a kid growing up in Brooklyn, virtually in the shadow of Ebbets Field, where the Brooklyn Dodgers played baseball (the fans, especially the youngsters, lived and died with them each summer), I had a bedside radio made by Motorola. With its aid, I listened to the marvelous southern tones of Red Barber announcing the Dodgers' games, either at home, where you could hear the roar of the fans, or as re-created by ticker tape reports when "Dem Bums," as the team was called, were playing in some far-off city like St. Louis. The name Motorola also evokes fond memories of other radio programs I listened to avidly after school or in the early evenings—"Jack Armstrong, the All-American Boy," "The Lone Ranger," "The Green Hornet," "The Shadow," and great comedy programs like "The Jack Benny Show," "Fibber McGee and Molly," and "Burns and Allen."

Well, Motorola doesn't make bedside radios anymore, but they certainly have come a long way in the telecommunications equipment business. And they almost certainly are going to continue that growth. They are now one of the world's leading manufacturers of electronic systems, equipment, and components. Their products include two-way radios, paging equipment, and a portable radio/data system that is probably the best in the world. They also are a large supplier of radio equipment to the armed forces, as well as of communications chips for switching used in satellites put up by NASA.

In March 1984, Motorola, Inc. announced a complete line of cellular portable and mobile telephones that are really outstanding pieces of equipment. If the cellular radio-telephone business takes off, as I think it will in many parts of this country, the Motorola equipment will be purchased and used in vast quantities.

Motorola, Inc. had an almost $800 million gain in net sales between 1981 and 1983 (it had sales and other revenues in 1983 of $4.33 billion, a gain of 14 percent), and in

1983 it had net earnings of $244 million, or $6.26 per share. Its return on average invested capital in that year was 12.2 percent, up from 9.1 percent in 1982. Through September 30, 1984, its sales exceeded $4 billion, and 1984 earnings are estimated to be $3.25 a share. Earnings in 1985 are projected at $3.50 per share. Most important to me was Motorola's heavy expenditure of capital on research and development, especially in what it calls its Communications Sector.

Under excellent top management, consisting of Robert W. Galvin, chairman, William J. Weisz, vice-chairman, and John F. Mitchell, its president, I expect big things from Motorola, Inc., and I think we'll see them.

MATSUSHITA AND PLESSEY

Not all of the best companies in telecommunications these days are based in the United States, but the securities of some of the best abroad are traded in the United States in the form of what are called American Depositary Shares (or Receipts). These are securities representing a certain number of shares of common stock in a foreign corporation, and they are registered with our Securities and Exchange Commission so that they may be offered to the public and traded as though they were the actual shares of the parent corporation. As the fortunes of the company increase, the value of the ADRs may as well.

The two outstanding examples of such companies are based on opposite sides of the globe, but both are very much involved in the telecommunications industry. From Japan, Matsushita Electric Industrial Company, Ltd. (NYSE) is truly one of the world's great companies, but the general public here only recognizes it by products using the names Panasonic, Technics, and Quasar, and in other countries the name National. It operates in more than 130 countries and produces some of the finest equipment made anywhere. It produces all sorts of audio and video equipment. You may have enjoyed the almost crystal-clear large color video display system produced by Panasonic for the 1984 Summer Olympic Games at the Los Angeles Coliseum.

In 1983, Matsushita had sales of almost $17 billion and net income of $774 million. Each of its American Depositary Receipts represents ten shares of common stock and, assuming full dilution, it earned $4.82 per American depositary share that year. Through the nine months ending August 31, 1984, Matsushita had net sales of about $14 billion (20 percent above last year), and their earnings per ADR as adjusted were $3.96. Its net income for the fiscal year ending November 30, 1984, is estimated to be almost $1 billion on sales of $19.5 billion.

Matsushita, unlike its great competitor, SONY Corporation, chose to place emphasis on the VHS videotape recorder system instead of the Beta system, and it has paid off handsomely. Sales of such VTRs by Matsushita reached $2.55 billion for the first half of 1984, 26 percent ahead of the previous year. Sales of video equipment as a whole rose 19 percent for Matsushita. It also grew by 42 percent in its 1984 first half sales of communication and industrial equipment, and overall sales of electronic components rose 58 percent in the first half over the comparable 1983 figure.

One fascinating fact about Matsushita is that approximately one-tenth of its 124,000 employees are engaged in research and development at twenty-two research laboratories, including the magnificent Central Research Labs supported by the Japanese government. The company spent $738 million, or 4.4 percent of total sales, on research and development in 1983, a marvelous way of making certain that the company's fortunes will continue to increase.

On the other side of the world, based in Essex, England, is the Plessey Company plc (NYSE), whose American Depositary Receipts (one for every ten common shares) are traded on the New York Stock Exchange and have been doing very well over the past couple of years. I know a little bit about Plessey because my former law partner, Warren J. Sinsheimer, is also deputy chief executive of the parent British company and chairman of its U.S. subsidiary, and I have done some work for Plessey here.

Plessey has been the largest supplier of telephone equipment in the United Kingdom and also supplies military materials in telecommunications, including radar, to the Ministry of Defence. The company has taken a position in the ownership of Scientific-Atlanta, Inc., an excellent U.S. company manufacturing communications satellite dishes and related equipment. Plessey is exploiting this new business in Europe and the Middle East.

Plessey's fiscal year ends on March 30. Its financial report for the year ended March 30, 1984, reveals its significant growth and even greater potential. It reports only in English pounds, but even with the decrease in the value of the pound against the dollar the figures are impressive. Profits before taxation were over £176 million, a 20.3 percent improvement from the previous year. This was based on total turnover of £1,218.9 million, an increase of 13.4 percent from the previous year, and operating profit went up by 22.9 percent to £146.3 million. Over a five-year period, Plessey increased its earnings per share from £5.57 in 1980 to £15.25 in 1984. Through September 1984, Plessey had net sales of over £1.1 billion, a rise of 5 percent over 1983, and operating profit was up by over 3 percent.

Again, Plessey is one of those companies that steadily have invested and built for the future in telecommunications. It has built a number of new manufacturing and technology centers over the last five years, and total investment in new buildings and the refurbishment of existing production and research facilities has exceeded £233 million in five years. Plessey's Allen Clark Research Centre at Caswell, England, has earned a worldwide reputation for its work on telecommunications materials, as well as for a number of new electronics applications. The Centre has also added a new high technology building, which is only the first phase in the new plan to provide additional facilities for Plessey's solid-state research.

Under the leadership of Sir John Clark, Plessey has indicated its intention to strengthen its position of industrial leadership in world electronics and communications, and I fully expect that such a plan will be fulfilled. Purchasers of its American Depositary Receipts should benefit from those results.

A. H. BELO

A company that has been extraordinarily active in the broadcasting field in the southern part of the United States, the so-called Sun Belt, is A. H. Belo Corporation (NYSE). In 1983 it acquired Dun & Bradstreet's television properties for $606 million. These former Corinthian stations have been doing quite well, and should add to the earnings of Belo's *Dallas Morning News*.

As debt is reduced, there should be some excellent cash flow in this company and significant profits by the end of the decade. A Drexel Burnham Lambert research report in mid-1984 raised the 1984 estimate of earnings for Belo from $2.60 a share to $2.80, while their 1985 estimate remained the same at $3.40 a share because of expected higher newsprint costs. Through September 30, 1984, revenues exceeded $200 million, up over 50 percent from the corresponding prior period.

The management of the company is strong and aggressive. It includes Joe M. Dealey as chairman, G. B. Dealey as president and chief executive officer, and Robert Dechard, the largest single stockholder in the company (only thirty-two years old). The head of its broadcasting operations is Ward L. Huey, Jr., who has been involved in the industry for a long time.

I would not be surprised to see A. H. Belo make a few

more television and radio acquisitions in the next year or so. They are going to be aggressive as the FCC rules change, and the company will almost certainly attempt to be a major force in the broadcasting field.

ADVANCED SYSTEMS

Perhaps one of the least-known public companies very much involved in the Information Age is Advanced Systems, Inc. (NYSE). This company used to be a subsidiary of URS Corporation, also a public company involved in the energy equipment field, but fortunately it was spun off a couple of years ago and now exists on its own, although some of the URS people, including Arthur Stromberg, the chairman, remain involved. ASI's shares are now traded on the New York Stock Exchange, but it hasn't been as fully appreciated in the market as I think may be warranted by its activities and results.

ASI originally was set up to develop and market video-cassette training materials, and it established world-wide distribution of materials it developed in Illinois. With the growth of computers, ASI has become very much involved in computer-assisted instruction and computer-based training, while it still offers a library of more than 3000 video learning modules in subjects such as Information Age technology management, manufacturing, home marketing and sales, electronics, engineering, human resource development, and data processing skills. Using video- and audiotapes as well as printed guides, texts, and workbooks, all sorts of training may be carried out. Combining laser and videodisc technology, ASI has established Interactive Video Instruction so that users may learn various skills at their own pace.

For its fiscal year ended October 31, 1984, ASI produced revenues of over $40 million but net income fell by about 28 percent. However, ASI's officers expect fiscal 1985 to be its most profitable year. Under the able leadership of William

R. Roach, its president and chief executive officer, ASI has moved to the forefront of the telecommunications training field, and I fully expect that it will continue its rapid growth. It is certainly one company whose stock price requires only a small investment of capital and is susceptible to significant increase.

CHYRON

With the growth of commercial and public television, cable television, pay cable television, and other video media, the need for equipment to produce programming has become tremendous. Chyron Corporation (OTC) is the leading manufacturer of high performance electronic digital graphics equipment for the broadcast and video production industries on a worldwide basis. It was one of the first companies to apply digital computer technology to generate titles and graphics electronically, and Chyron-developed state-of-the-art equipment and software for the television broadcasting industry is known everywhere for its high quality, reliable performance, and cost effectiveness.

Essentially, Chyron makes the equipment that creates all of the letters, numbers, titles, and logos that appear on television programming, including the news and sports programs, advertising commercials, and even some of the reports on the stock market. Their equipment is used by the three major networks in the United States and is so highly regarded that it is also being used in many countries abroad. It has been estimated that Chyron has 70 percent of the market for basic graphics generators for the broadcast industry, with equipment ranging in price from $19,900 to $35,000.

While Chyron has grown rapidly, it has done so in a very orderly manner. It has constantly expanded and upgraded its existing products and technology, and it has moved new versions of existing products into new markets and sought new uses for existing products. It is also very much involved in research and development, and it has the cash (over $17 million at the end of its 1984 fiscal year) to engage in such research and development.

Its stock has been generally undervalued by the market. If it continues the growth pattern evidenced by its consistent increase of net sales, Chyron certainly is going to be one of the best buys in the telecommunications field today.

LIN BROADCASTING, TAFT BROADCASTING, COX COMMUNICATIONS, AND MALRITE COMMUNICATIONS

One of the nicest things that has happened in the broadcasting industry in the United States from an economic point of view was the appointment of Mark Fowler as chairman of the Federal Communications Commission (FCC) in 1981. Mr. Fowler has made it his basic objective to deregulate the telecommunications business wherever possible, particularly in radio and television. One recent effort at such deregulation may very well have a significant beneficial result with respect to certain companies that own and operate radio and television stations in the United States. Although there is a general fear that the network corporations want to own and control a large number of broadcasting stations so that they can control programming sold to such stations, I believe that this will not occur.

Instead, the result of recent deregulation changing multiple-ownership rules, with respect to the number of stations that one entity may own, will be that certain companies presently in the field will be able to own a larger number of stations, operate them in a more businesslike manner, and produce greater cash flows and profitability. Recently the rules were changed so that now any one person or entity can own up to twelve AM and twelve FM radio stations. There is still controversy on the rule on multiple ownership of television stations. In 1990, assuming that there are no problems with such multiple ownership, there may be no restrictions at all.

The companies that are best geared to make further acquisitions to effectively enlarge their number of owned and operated stations (besides the aforementioned Capital Cities Communications, Inc.) are, in my judgment, LIN Broadcasting Corporation (OTC), Taft Broadcasting Company (NYSE), Cox Communications, Inc. (NYSE), and Malrite Communications Group, Inc. (OTC). Interestingly enough, all of these companies have also diversified in one way or another in the telecommunications field, and their future profitability may be enhanced by that diversification as well as by expansion in radio and/or television.

LIN Broadcasting Corporation owns ten radio stations and seven television stations, including two television stations purchased in 1984 for about $105 million. All of them are in generally excellent markets. LIN has also gone into the radio common carrier and cellular radio-telephone businesses, including a 45 percent joint venture position in a partnership that has been awarded a cellular permit for the New York City area. I expect the company will do quite well in both of these fields. LIN also owns 51 percent of a cellular radio-telephone company operating in Philadelphia, Pennsylvania, and 35 percent of such a company operating in Los Angeles.

In 1983, LIN had net revenues of $107.333 million and net income of $23.528 million, or $1.04 on a fully diluted basis. At the end of 1983 it also had working capital of $138 million and total assets of $296.345 million. Their long-term debt was only approximately $106 million. In the first half of 1984, LIN showed net revenues of almost $70 million, well ahead of 1983. Its common share earnings have grown consistently from 80 cents in 1981 to 93 cents in 1982 to $1.06 in 1983, and the trend seems to be continuing.

Under the very able leadership of Donald A. Pels, the chairman of the board and president of LIN, the company is well prepared to carry out significant growth in ownership and operation of more radio and television stations. I also expect that it will continue its conservative but persistent entry into the radio common carrier and cellular radiotelephone operations, and these ultimately will be quite successful for LIN.

Since 1939, Taft Broadcasting Company has been involved in the telecommunications business, first in radio and then in television. In recent years, it also has become involved in development and production of television programming, on both an animated and live basis, and it also has become a distributor of programming. Taft has additionally gone into the cable television system ownership business (with Tele-Communications, Inc. having almost 150,000 basic subscribers in New England and Michigan), and it has made some small investments in cellular radiotelephone, teletext, video conferencing, and multipoint distribution service, besides obtaining the license to operate a low-power television station in Wisconsin.

Taft Broadcasting Company has seven television stations in some of the largest markets in the United States, and has thirteen radio stations in certain of the same markets, as well as in key cities like Tampa–St. Petersburg. Taft increased its net revenues from $167.1 million in 1979 to $453.8 million in the fiscal year ended March 31, 1984. Its earnings from continuing operations also increased from $25.9 million in 1979 to $39.2 million in 1984, with earnings per share from continuing operations growing from $3.06 to $4.03 during the same period.

I look for Taft to be one of the leaders in acquiring and managing additional broadcasting operations, thus creating an increase in its long-term value.

Probably the most dynamic group broadcasting company in the United States today is Cox Communications, Inc., based in Atlanta, Georgia. Cox Communications owns television stations in cities like Atlanta, Charlotte, North Carolina, San Francisco–Oakland, Chicago, St. Louis, Pittsburgh, and Dayton, Ohio. It also owns twelve radio stations in most of those cities, as well as in Miami and Philadelphia. Its Cable Communications division now has fifty-four systems in twenty-three states, including the largest system in the United States, located in San Diego, California. Its systems serve over one and a half million basic subscribers.

The revenues produced by Cox's Broadcasting Division increased from $133.7 million in 1979 to $231.5 million in 1983. The company had an operating income in 1983 of $87.1 million from its broadcasting operations alone. Overall, in 1983 Cox Communications, Inc. had net revenues of $614.623 million, a 19.4 percent increase from its 1982 net revenues, and it had a 19.4 percent increase in operating income, rising from $118.175 million in 1982 to $141.082 million in 1983. Its net income per common share went from $2.31 to $2.75. Results for the nine-month period ending September 30, 1984, indicated that 1984 was another record year for Cox. Revenues increased 20 percent and operating income was up over 19 percent over a year earlier. Its cable division reported increases in revenues and in operating income. Cox has participated in franchise applications for building cable systems in parts of New York City, and it probably has been the most conservative of all of the applicants there. I expect that if anyone makes money in cable in New York City, it probably will be Cox Cable.

One of the more recent entrants into public ownership of companies in the broadcasting industry is Malrite Communications Group, Inc., which made an offering of some of its common stock in November 1983 at $16 per share. Malrite, a Cleveland-based company led in a most dynamic fashion by Milton Maltz, owns and operates three independent television stations and approximately fourteen AM and FM radio stations. At the present time, Malrite is in the process of selling some of its existing stations and purchasing additional ones.

As an example of its aggressiveness, Malrite purchased an insignificant FM radio station operating out of Newark, New Jersey, that was broadcasting "beautiful" music and which had obtained FCC approval to move its antenna to the Empire State Building in New York City. Upon purchase, Malrite changed the call letters to WHTZ (FM) and changed its format to what is called "hot hits." That station, generally referred to as "Z-100," almost immediately went to first place in the New York City ratings, and it has become a popular and very profitable operation.

Malrite has indicated that it will continue to upgrade its broadcasting properties and will move into even larger radio markets. With the change in the multiple-ownership rule, I expect that Malrite will use its continuing cash flow and the proceeds from its sale of stock to the public ($32.56 million) to enlarge its group and produce continuing profit.

VIACOM

One of the best investments around, Viacom International, Inc. (NYSE) was organized in August 1970 as a wholly owned subsidiary of CBS, Inc. to run the television program distribution and cable television businesses of CBS. In June 1971, pursuant to government direction, CBS transferred these businesses to Viacom and distributed its shareholdings in Viacom to the holders of CBS common stock on a *pro rata* basis.

While its principal business continues to be television program distribution and cable television system ownership and operation, Viacom has also gone into the television production business and owns 23 percent of the pay cable television networks known as "Showtime" and "The Movie Channel." In recent years, Viacom has acquired four television stations, and it operates five FM and one AM radio stations.

Viacom's revenues increased in ten years from over $38 million in 1974 to over $315 million in 1983. Its net earnings also increased, from \$3.044 million in 1974 to \$28.129 million in 1983, and on a fully diluted basis, its net earnings per common share have increased from 36 cents to \$2.05 over the same period. Its return on an average shareholder's equity in 1983 was 13.7 percent. Revenues for the nine months ending September 30, 1984, dropped by about 5 percent, but profitability increased.

While Viacom, under the outstanding leadership of Chairman Ralph M. Baruch and President and Chief Executive Officer Terrence A. Elkes, has not done everything right in positioning itself for future success in the telecommunications and entertainment fields, it has done a lot of things right. Viacom does very well in the marketing and distribution of television programming, including the number one first-run syndicated television program "Family Feud." The ownership and operation of cable systems through Viacom Cable have been carried out in a somewhat conservative fashion in such places as San Francisco, California, and Nashville, Tennessee. Its systems have a total of 720,000 subscribers, with almost 600,000 pay units and 11,000 miles of cable distribution. It has wisely avoided making applications for franchises in very large urban areas, and it hasn't exhausted its resources in attempting wildly to go after large franchises. The company stayed with the essential requirements of cable systems in areas that should continue to produce significant cash flows and profitability.

Viacom's recent acquisition of television stations in Rochester, New York, and Shreveport, Louisiana, as well as its persistent desire to become more involved in ownership of radio stations, indicates a fine overall game plan in preparation for operating a large group of stations under the new rules.

Ralph Baruch's 1984 decision to relinquish the chief executive post to Terry Elkes doesn't in any way mean that he won't remain significantly involved in the management of Viacom. It does indicate the seriousness that Viacom attaches to both succession planning and promotion from within; they have an excellent group of upper and middle management executives to continue the company in its positive direction for many years to come.

LORIMAR

One of the prime areas for growing revenues in the telecommunications field in all parts of the world is television program syndication. One of the best companies involved in syndication is Lorimar Corporation (ASE), which has done extraordinarily well and will do even better. With its expanding syndication of one of the most successful television programs in the world, "Dallas," Lorimar should have earnings in 1985 that double those of the prior year.

Revenues for the fiscal year ending July 28, 1984, were 43 percent higher than for the prior year, and Lorimar's earnings per share for the fiscal year ended July 31, 1984, were $2.06. This included almost $44 million of revenues from the Kenyon & Eckhardt advertising agency, which Lorimar acquired in 1984. K&E is one of the biggest agencies, with some very talented people in the television field, and under Lorimar's management it should also do better. In November 1984, Lorimar acquired Karl Video Corporation, which produces and distributes home video programming.

Merv Adelson is chairman of the board and Lee Rich is president of Lorimar. Both are very talented and experienced men in the television business, and they have some fine people surrounding them. To me it seems certain that Lorimar will be one of the supreme suppliers of software in the telecommunications field throughout this decade, and the market value of the company's shares will reflect the already evident world-wide growth of television viewing.

PAGEAMERICA

If you want to make a lot of money, sometimes you have to be a little less conservative in choosing public companies to invest in. I've tried here to exercise my usual caution, but I would like to mention a company that's involved in a more recent aspect of telecommunications that I believe will be highly profitable in the years to come. The company is now called PageAmerica Group, Inc. (OTC), and it's in the radio paging products and services business in the United States.

At one time, David A. Post, PageAmerica's chairman of the board and chief executive officer, was a client of my former law firm, and for a while he was having great difficulty trying to persuade people to back him in establishing PageAmerica. However, he's a most persistent and ambitious young man, and he has succeeded in obtaining equity financing from major institutional investors in excess of $12 million to further the activities of PageAmerica. David fortunately brought in Steven L. Sinn as president and chief operating officer and has also brought an excellent management team on board.

In two years, PageAmerica has grown from 2500 radio paging subscribers to over 30,000 internally—and to 50,000 upon completion of some acquisitions. It is now reportedly the eighth largest radio common carrier in the country. PageAmerica, in partnership with RCA Global Communications, Inc., has developed and operates a worldwide paging/message/information network. The services are being offered in several cities in the United States, and they include "pocket electronic mail." It's expected that the radio paging industry will grow to at least 20 million subscribers by 1990, utilizing additional new frequencies authorized by the FCC, as well as new paging technology and newly established paging networks.

PageAmerica's products and services include "PageGram," which uses a small telecommunications device worn by the subscriber that displays the telephone number of the calling party, and "PageTone," which is a tone-only device permitting two different calling sources to contact the wearer. Both operate by radio signals sent from a central office over a certain frequency to the subscriber carrying either device. PageAmerica also expects shortly to offer an alphanumeric PageGram that will receive and store text messages as well as numbers.

Through the utilization of communication satellites, in a few years companies like PageAmerica will be able to achieve instant delivery of messages and information to vehicles and fixed terminals. Pocket-size receivers with a tiny screen that prints out a message will become standard items for many business people and professionals. Messages can then be delivered to them wherever they are. As PageAmerica indicates in its own materials, what this means is that any individual or group will be able to send information to other individuals or groups instantly, no matter where the recipient is located. The company plans for the system to be remarkably simple and inexpensive to use.

While there will be extraordinarily severe competition in the paging business in the coming years, I think that PageAmerica has positioned itself to be one of the leading companies in the industry as development takes place. Their stock is priced advantageously these days to make appreciable profits possible as that company prospers.

TELE-COMMUNICATIONS, INC. AND
ROGERS CABLESYSTEMS

I know that there has been a very bearish view taken by investors toward the entire cable television field in the past two years or so, but this has been caused by two factors that have nothing to do with the industry as a whole. Because of such unwarranted pessimism about cable television companies, this is a good time to buy some of the depressed cable stocks—before too many people wake up to the fallacies that have been bandied about concerning the industry as a whole.

The unfortunate fact is that it took a long time for most

people to realize that cable television was really a fantastic industry and that being a cable systems operator was very lucrative. When I was writing the first edition of *All About Cable* in 1979, I found that almost everyone thought of cable television as an industry that had just been born, rather than as one that was already penetrating almost 25 percent of the television households in the United States.

When everyone started getting on the cable bandwagon, too many thought that the big money would be made in starting a satellite programming network. Many people and companies believed that they could create instant television networks by renting transponders on communications satellites (at very high monthly costs), buying or producing programming geared to special interests (again very expensively), and marketing their networks to the cable system operators, who they thought had to appeal to the diverse interests of existing or potential subscribers. These satellite network enthusiasts were sure cable system operators would either accept their programming for a nominal monthly, per-subscriber fee or take the programming without charge but let the network operator sell a lot of advertising to companies just dying to place advertising on their channels.

Everyone wanted to get in on the act, from CBS Cable and the Entertainment Channel (in which Rockefeller Center, Inc. and the former president of CBS, Inc., Arthur Taylor, were involved) with their cultural programming networks, to a health channel (backed by Viacom International, Inc.) and one called Cinemerica, that was supposed to have at least twelve hours of programming a day geared to people fifty and over.

There was a proliferation of publicity on how Company X was leasing a satellite transponder for millions of dollars to deliver "new" kinds of programming and how Company Y was going to deliver three channels of programs each day to cable systems and make millions of dollars in six months.

The *programming* people became identified with the actual cable system owners, who have been operating old systems quite profitably all along.

When the cable *programming* people failed, as even CBS, Inc. did, the smart money assumed that the whole cable industry was going to fail, too, notwithstanding the tremendous success of the cable operators and even of operations like Home Box Office that deliver programming to cable subscribers by satellite. Perhaps this failure alone still would not have caused the resulting bearishness in regard to the entire cable industry. Combined with it, however, was the sorry story of Warner-Amex Cable Communications, Inc., a company owned jointly by Warner Communications, Inc. and the American Express Company. It overextended itself in spending millions of dollars in an experiment called "QUBE," prematurely trying to prove that two-way, interactive cable could be economically viable, before cable itself had been fully proven. It also filed applications for cable system franchises in many of the largest cities in the United States, including parts of New York City, offering extensive services at a cost the company could never cover, especially during a recessionary period, even while one of Warner's then-affiliated companies, Atari Computer, was in such sad shape.

Notwithstanding the misunderstood view of the cable television industry, I'm certain that it will go on to be one of the world's leading businesses by 1990. Today, several companies are already profitable and in a posture to create vast earnings in the cable business. One of these is Tele-Communications, Inc. (OTC). TCI, based in Denver, Colorado, is the largest cable television system operator in the United States, with more than 2.766 million subscribers. In 1983, TCI had revenues of $347.2 million, and operating income of $84.5 million. Its earnings per share were 46 cents, twice what they were in 1982. It appears that 1984 was equally profitable for TCI.

One of the nice things about TCI, aside from its excellent management, led by Chairman Bob Magness and President John C. Malone, is that it essentially has stayed out of the large urban areas so has not had to spend vast sums building its plant and placing its equipment in the cities. However, it is in some cities, and recently agreed to purchase the cable franchise in Pittsburgh from Warner-Amex. TCI certainly has the capability to continue doing well. As they keep expanding their operations, this should be reflected in their common share value and price.

Rogers Cablesystems, Inc. (ASE) is another of those "foreign" companies with securities being traded in the stock markets in the United States. Rogers is based in Toronto, Canada, and owns and operates cable television systems in Canada quite successfully. Through its American subsidiary, Rogers U.S. Cablesystems, Inc., the company operates cable systems in such cities as Minneapolis, Minnesota; Syracuse, New York; Portland, Oregon; and San Antonio, Texas; and in other areas that include southern California.

In 1983, Rogers had revenues of $364.2 million, with a cash flow from operations of over $48 million. They did even better in 1984 and will continue such growth for a number of years to come, even with the unfortunate decline of the Canadian dollar against the U.S. dollar. Modestly priced, Rogers Cablesystems, Inc. securities may be one of the best buys to make in the cable field.

ABC AND CBS

What's going to happen to the big television network companies with all this new technology in telecommunications? Aren't they going to die an uneasy death? Those questions get asked whenever the subject of the new media comes up. I've heard them hundreds of times. My response is always that you have to look at each of the network companies as a separate entity to understand them.

You certainly have to look at the National Broadcasting Company that way, because it's only one part of the vast RCA Corporation, which I previously discussed. While it functions in the same way essentially as the ABC and CBS radio and television network divisions, NBC is not a separate corporation. Many people are surprised to learn that the networks don't own all of the television and radio stations that carry their programs and the advertising that they sell nationally. The networks are permitted by FCC regulation to own only a limited number of stations themselves. The rest of the carriers of their programs are affiliates, independently owned licensees that are *paid* by the networks to carry their programs and their commercials.

Naturally, the proliferation of programs being fed to the public via various new technologies, including cable, SMATV, MDS, DBS, and videocassettes and videodiscs, has had an effect on the networks and even independent broadcasters. However, it seems to me that the effect will not be disastrous if the networks take advantage of their great capabilities, especially in the news and sports areas in which they are so capable and for which they are able to attract the best personnel. This doesn't mean they can't hold their own in providing entertainment, even cultural programming, but they'll have to be much more creative and aware of the changing tastes of the fickle public in order to maintain their lofty position and profitability with respect to advertising on such programs in the years to come.

American Broadcasting Companies, Inc. (NYSE) is and will remain one of the most outstanding companies in the world in the telecommunications field. It's a diversified company. Its operations include not only the ABC Television Network, but also ABC-owned AM and FM radio and television stations, five radio networks, magazines, television program production and distribution operations, motion pictures, and scenic attractions. It has also put its toes into cable television, fortunately in a very low-key manner.

One development has been purchase of an 85 percent interest in ESPN, the cable sports programming network.

ABC has made some mistakes, too, in going into new areas of telecommunications, but its management (with a great assist from its Research Services division) has been wise enough to commit only a limited amount of resources to those ventures and to admit their mistakes. An example of this was TeleFirst, by which ABC Video Enterprises sought to deliver video programming in scrambled form for automatic recording and subsequent playback by videocassette tape recorders in subscribers' homes. ABC apparently realized too late that most people really don't know how to work the timing devices in their videocassette recorders and didn't want to set them to start recording programming at 2 A.M. In any event, ABC ended this experiment in June 1984, at an expected cost to its pretax earnings of about $15 million. In spite of this, ABC probably will have had an excellent year in 1984, especially with the very fine job done in broadcasting both the Winter and Summer Olympics in 1984.

In 1983, American Broadcasting Companies, Inc. had total revenues of $2.9 billion, an increase of almost $300 million from its 1982 revenues. Its earnings per common share in 1983 were $5.45, slightly lower than the $5.54 per share in 1982. However, ABC will probably show earnings of about $6.70 per common share in 1984.

ABC Video Enterprises is also involved in cable programming services that haven't as yet proven profitable. However, one of their operations, the Arts and Entertainment Network, which ABC operates with the Hearst Corporation, projects that eventually some 9 million subscribers in cable systems will take ARTS' full range of American and internationally produced programs devoted to the visual and performing arts, in addition to certain entertainment programs from the British Broadcasting Corporation. With its acquisition of 85 percent of ESPN, ABC is now in a posture to deliver significant amounts of programming to cable subscribers with advertising support that may very well prove extraordinarily profitable—if the company can hold out that long—when most of the cable systems throughout the United States are built and such programming can be delivered on a wide scale.

Whether or not the ABC Television Network as a programming entity will continue to be as strong economically as it has been in recent years remains a question. But as I've indicated, the news, sports, and information aspects of such programming will almost certainly continue to be viewed on a large scale.

For long-term growth in the programming aspects of telecommunications, American Broadcasting Companies, Inc., under the continuing guidance of Leonard H. Goldenson, its chairman of the board and chief executive officer, and Frederick S. Pierce, who took over as president and chief operating officer of the corporation in January 1983 (and was recently given a new long-term contract), will probably be one of the wisest investments you can make in the field.

What can you say about a company like CBS, Inc. (NYSE), which has made a number of mistakes in certain telecommunications operations and still keeps increasing its net sales and common share earnings? While CBS has changed presidents on a number of occasions over the last few years, fortunately William S. Paley, its founder, is still active in its planning. Also, it has an outstanding corporate management committee, which has seen it through some of its more unfortunate days and is planning for sustained growth in the field.

While continuing to be a major factor in the programming area through its CBS Television Network, CBS also operates television and radio stations in the major markets of the United States, and it is very much involved in the record business (one of its subsidiaries has Michael Jackson under contract), as well as consumer products, publishing, and toys. New aspects of its operations are the production of medical educational materials and, through the acquisition of Transmedica, Inc. in May 1984, the provision of a national cable television program service for physicians. The professional training area could provide some very interesting profits for CBS in the years to come.

CBS is not too dissimilar from ABC. Its net sales in 1983 were $4.458 billion, almost twice as much as ABC. Its profits from its broadcast operations in 1983 were $291.5 million, or 53 percent of CBS's total revenues that year. In all likelihood, the company exceeded such earnings in 1984, and it is expected that its common share earnings may have increased from $6.31 per share in 1983 to an estimated $8.90 for 1984. Its operating income in 1983 was $398 million, almost $100 million more than in 1982.

CBS now participates in the very successful CBS/20th Century-Fox home videocassette operation, now the largest in the home video business. It's also a participant, with Columbia Pictures and Home Box Office, in Tri-Star Pictures, which is now producing theatrical films such as *The Natural* and *The Evil That Men Do*.

CBS also has formed an Operations and Engineering Division. This division designs and implements technical installations at all CBS television and radio facilities, and it also coordinates the ongoing development of new technologies. I suspect that this division will shortly be facilitating the complete delivery of CBS television programs to the network's owned and operated stations and its affiliates via communication satellites, which will save the company a good deal of expense in leasing land lines. I also expect that CBS will be delivering television programming in stereo soon, and that the company's production capabilities will be utilized for such stereo programming by mid-1985.

It's quite apparent that Mr. Paley has placed the power of management of CBS, Inc. in the hands of Thomas H.

Wyman, its chairman, president, and chief executive officer, with a view to having him carry out an established game plan for taking CBS, Inc. fully into the Information Age in a meaningful way. With the other officers of the various groups that CBS has established contributing to this administration's plans and operations, I wouldn't be surprised to see CBS stock at $100 a share in the not-too-distant future.

TELEPICTURES CORPORATION

While by no means an ABC or CBS, Telepictures Corporation (OTC), a company organized only in September 1978, appears to me well worth looking into carefully for investment. Telepictures is engaged in marketing and distributing films and programs primarily to television stations, is involved in the production of television films and programs, and owns television stations. Telepictures originally made its mark in distributing old "I Love Lucy" television programs in various parts of the world; this proved to be enormously profitable. In 1983, 31 percent of its revenues were derived from the sale of the television series "More Real People," and 20 percent was derived from the first-run syndication of the very popular program "The People's Court."

Telepictures has also purchased other companies involved in television production activities, and it develops and distributes a feature news service—News Information Weekly Service—with materials sold to television stations for inclusion in the local stations' news broadcasts.

The company acquired KMID-TV, in Midland/Odessa, Texas, and in 1984 it agreed to acquire two inactive television stations in Puerto Rico that have been involved in bankruptcy proceedings. Those Spanish-language stations could well prove the forerunner for Telepictures' acquisition and operation of other such stations in various parts of the United States and elsewhere.

Telepictures had $71.1 million in revenues in 1983, with operating income of $10.5 million. In 1984, it probably produced twice as much revenue. In December 1984, Telepictures Corporation's common stock was selling for approximately $16 per share. If the company keeps up its expansion of activities, it should certainly maintain the growth in its revenues and earnings and in its share price.

SCIENTIFIC-ATLANTA

Probably the best "buy" of all the securities mentioned here is Scientific-Atlanta, Inc. (NYSE), led by Sidney Topol, a dynamic fellow with great foresight in the telecommunications industry. This Atlanta-based company was one of the first to get into manufacturing equipment for satellite communications. It had some tough times during the 1982-83 recession, but it has come through and its earnings in fiscal 1983-84 (its fiscal year ends June 30) certainly indicates such success. It has had a tremendous order backlog, and I don't think it has even begun to scratch the surface of the need for satellite communications equipment it can supply throughout the world. As previously indicated, it has an agreement with the Plessey Company, giving it a strong foothold in the United Kingdom and Middle East markets, as well as in Australia and Africa.

Scientific-Atlanta has also entered into the cable television addressable set-top terminal business, and this apparently has proven a successful venture. The company's equipment is being used by Tele-Communications, Inc. and Falcon Communications (which recently bought a number of cable systems from Warner-Amex Cable Communications).

In its fiscal year ended June 30, 1984, Scientific-Atlanta, Inc. had net sales of $398.9 million. Its common stock earnings have gone from 90 cents to 63 cents to 20 cents to approximately 50 cents over a four-year period; the trend is obviously upward again. Sales growth is expected to increase in the 1984-85 fiscal year.

Scientific-Atlanta is providing earth stations for satellite distribution programming being transmitted by CBS Television to its owned and affiliated stations. Delivery of this equipment is supposed to be finished by the middle of 1986.

I expect that the introduction of John H. Levergood as chief operating officer, the association with Plessey, which gives Scientific-Atlanta access to Plessey's computer and software systems, and the tremendous ongoing demand for satellite and cable equipment will place Scientific-Atlanta, Inc. in a leading position in the telecommunications equipment industry by 1990. I was surprised to find that its common stock was selling for less than $10 a share in 1984, but that probably provides the investor who wishes to multiply his investment rapidly an opportunity to do so.

GTE

Many people in the United States know GTE Corporation because that company has long been providing basic telephone service in their areas. Now, many other people know GTE because of the expansion of its "Sprint" long distance telephone service, especially with the advent of equal access for all to such services. It clearly is one of the leading telecommunications companies in the United States, and with the breakup of AT&T and other factors, GTE Corporation (NYSE) is probably going to continue its growth to become one of the leading companies in the world.

In 1983, GTE's overall revenues and sales totaled almost $13 billion. From its telephone operations alone, it had total revenues of $8.3 billion. Its overall net income totaled $926 million, and its common share earnings in 1983 were $5.12 per share. These figures have been increasing strongly in 1984. The simple fact that more people are using telephones these days and will continue to do so, plus the fact that the company has been able to increase its rates, indicates that GTE's telephone revenues and earnings will continue to rise in 1984 and beyond.

GTE is very much committed to technology, research, and development. In 1983, it spent approximately $262 million for company-sponsored research and development programs carried out by 2900 scientists and engineers in the company. In 1984, this expenditure reached almost $300 million.

It seems strange that a company as large and dynamic as GTE Corporation should have had a book value per share at the end of 1983 of only $33.89. At the end of 1984 its common stock was selling on the New York Stock Exchange for approximately $40 per share. As the telecommunications industry continues to increase and as GTE keeps growing, I look for its shares to sell for considerably more than their book value. As they say in their television commercials, "Gee!—No, GTE."

THE TRIBUNE COMPANY

It cannot be purely coincidental that the Chicago Cubs baseball team finally has risen to some heights at the same time that its owner, the Tribune Company (NYSE), has become a diversified communications public company with operating revenues in 1983 of over $1.5 billion. The Tribune Company owns more than the Cubs; it publishes Chicago's newspaper institution, *The Tribune*, as well as the *Daily News* in New York City, the *Los Angeles Daily News*, the *Ft. Lauderdale News and Sun-Sentinal*, and other newspapers. It also owns a number of television stations, including WGN-TV in Chicago and WPIX-TV in New York, and it's the owner of four radio stations and a number of cable television systems.

One of the best cable franchises in the United States, in Tampa, Florida, is owned by the Tribune Company. Tribune's cable television operations consist of seventeen other systems in eleven states, serving over 145,000 basic subscribers. Tribune has purchased additional television stations in New Orleans and Atlanta, and I expect that it will continue to make acquisitions in the broadcasting field to carry it to whatever the FCC's multiple-ownership rules may allow.

Tribune is also in the television production and distribution business and has played a leading role in Operation Prime Time, a consortium of independent television broadcasters that has developed some very fine television programming played on a network of independent stations. The Tribune Entertainment Company also produces and syndicates "At the Movies," a very popular weekly television program in which Chicago movie columnists Gene Siskel and Roger Ebert review current movies. The program is now carried by over 144 television stations throughout the United States each week. It started on public television.

The Tribune Company was privately held until October 1983, when it made its initial public offering and sold 7.7 million common shares at $26.75 each. Part of the proceeds was paid to the company, and part to private shareholders. The company's proceeds of approximately $125 million were used to repay about $70 million of debt and to finance the acquisition of the television station in New Orleans. The Tribune Company's common stock has not varied much in price on the New York Stock Exchange since that original offering, but if management continues to carry out its apparent business plan to build a large and diversified communications company, I believe that the stock price will multiply many times over in the next few years.

NYNEX CORPORATION

Probably the most important event in the history of telecommunications in the United States, if not the world, since Alexander Graham Bell called out to Mr. Watson for his help on the telephone, was the breakup of AT&T. In 1982 United States District Court Judge Harold Greene issued his modified final judgment calling for the divestiture by American Telephone & Telegraph Company of its twenty-two local telephone companies and the establishment of regional holding companies. The divestiture process called for separating out the local telephone exchange functions from the intercity telephone services and manufacturing operations of the Bell System, and the Bell operating companies were regrouped under the ownership of seven regional holding companies.

These regional holding companies, through their own subsidiaries (which were the Bell operating companies), are providing local telephone service. They are now also actively engaged in exploring opportunities to utilize new technologies and penetrate new markets for telecommunications services in order to maximize the value of their shares for their shareholders. One of these opportunities, cellular radio-telephone service, has become a reality for NYNEX Corporation (NYSE), one of the newly created RHCs, which owns and operates two of the most lucrative former local telephone companies, New York Telephone and New England Telephone. This entity now controls a major part of the Northeast. It offers service to the largest city in the United States, New York, as well as to one of the most advanced areas of high technology in the United States, the metropolitan area of Boston, Massachusetts. Assuming appropriate permission is given, these markets should enable NYNEX to enjoy significant revenues and earnings. In the first quarter of 1984 alone, NYNEX had revenues of $2.3 billion, and pretax income was almost $370 million. Indications are that it had an overwhelming profit in 1984.

NYNEX is one of the telephone industry's leaders in the use of fiber optics. It already has stated a plan to place more optical fiber in the New York City and Boston metropolitan areas to be utilized for the extraordinary data transmission needs of business operations there, as well as for expanded residential use. By the end of 1986, NYNEX estimates that it will have about 91,000 miles of fiber optics in place. This means that extraordinary amounts of digital information can be carried in these areas, and the company will also add more electronic, stored-program-control switching centers utilizing digital equipment.

While there is no certainty that this will mean continuing profitability for NYNEX, especially since the state regulators may not allow the kind of rates that should be required, I strongly suspect that the economic viability of the regional holding companies will be proved and appropriate cooperation will be obtained. This should be reflected in dividends to shareholders and in the value of the market price of the company's shares.

MCI COMMUNICATIONS

1984 has become the year of "equal access" in the telecommunications business in the United States, and it will be interesting to see how this affects one of the fastest-growing companies in the telecommunications business, MCI Communications Corporation (OTC). By regulation, the various telephone companies now must allow all companies offering long distance services equal access to the local exchanges in function and quality. In this context, MCI has estimated that it will, during the transition to full equal access, have monthly charges of about $330 per line, before giving effect to message unit credits from the local telephone companies that are expected to average about $20 per line. This means that during the time it takes to let all of AT&T's former long distance customers choose which new long distance service they want (or the customer may continue with AT&T service), MCI will have to charge some fairly high rates, but still probably lower than AT&T and certain other services.

William G. McGowan, the extraordinarily dynamic and brilliant chairman and chief executive officer of MCI, has spent millions of dollars in legal fees alone to bring MCI to the position that it now occupies in the long distance telephone company business. He has been assisted in an outstanding manner by probably the most intelligent commissioner ever to have served on the Federal Communications Commission, Kenneth Cox, who in the early days was general counsel for MCI and is now its senior vice-president. They have won many a battle, including a $1.8 billion judgment against AT&T that is still being challenged. It remains to be seen whether or not the regulatory position they have brought themselves to within the long distance telephone field (MCI still has only a small segment of the entire business) will remain profitable. However, MCI now provides a full spectrum of telecommunications services worldwide, and I believe it has the capacity to build an even greater, profitable business for the rest of the decade.

Through the use of microwave technology, fiber optics, and satellite communications—all utilizing the latest digital, tandem, and packet switches—MCI has been building a communications highway to transmit voice, data, and images throughout the United States. It already has invested over $2 billion in its coast-to-coast communications system and has nearly 17,000 route miles of high-capacity transmission facilities. It's the second largest microwave system in the country.

During its fiscal year ended March 31, 1984, MCI spent almost $900 million to expand its communications system and expects to invest about a billion dollars in the year that will end March 31, 1985. By the end of that fiscal year, MCI expects that its circuit miles will total almost 275 million (the number of miles that its equipment permits messages to criss-cross throughout the United States and the circuits that it sets up within each city that it services) and that it will have a nationwide, high-capacity network of more than 27,000 route miles (the number of actual miles that its microwave equipment carries messages between various cities throughout the country).

MCI has also entered the cellular radio-telephone business, and it will be providing such service in cities like Minneapolis–St. Paul, Pittsburgh, Denver, Boulder, and Cleveland. It will probably also offer such service in the Los Angeles area. Although this business will be competitive, MCI should have a good share of those hopefully profitable markets.

MCI also recently introduced a new service, MCI Mail, a type of electronic mail that can be used with any electronic communications device and will deliver a message to anyone anyplace either on a computer or on paper at prices that MCI claims are as much as 90 percent lower than similar services. I'm quite sure that this service won't be an overnight success, mostly because it takes a while for people to drop their old habits and adopt new ones. But in the long run, as MCI enlarges its communications capacity, particularly in the United States, and if it properly markets the service, that will become a sizable business with significant earnings.

Bill McGowan doesn't run the whole show at MCI anymore. He is now joined by V. Orville Wright, president and chief operating officer, and about 8500 other employees. With the divestiture of AT&T and economic growth throughout the world, MCI has a good chance to be one of the leading companies in the telecommunications field for a long time to come. I include MCI in this list of "best buys," but probably with the greatest hesitation of all of the companies discussed here, as there are so many uncertainties about its competitive position in the field.

AT&T

AT&T (NYSE) is still one of the greatest companies in the world, and I expect that it will continue in that position. At the time of divestiture, AT&T Communications serviced approximately 80 million residence customers and 7 million business customers. It provides regular AT&T long distance service, the Wide Area Telecommunications Services

("WATS"), and others, including satellite data and broadcast services.

AT&T also has a Technologies Division, which includes the Bell Laboratories, perhaps the best of all product development groups in communications science and technology since it was founded in 1924. It certainly seems reasonable to believe that the AT&T Bell Laboratories operation will continue to make great advances in telecommunications technology, which will then be profitably marketed by AT&T over the years to come.

I'm sure that almost all of us have had a relative who believed that buying and keeping AT&T shares (or bonds) meant financial stability for a lifetime. I believe this will be the case with the new AT&T, too, and was very surprised to see the shares of the company go down slightly after divestiture. By the end of 1984, the common stock of AT&T (it was still under $20 per share) was one of the biggest bargains around anywhere. It is only a small portent of the future that lies ahead for AT&T that for the nine months ended September 30, 1984, it had total revenues of over $24 billion and pretax net income of about $1.4 billion.

Since 1930 the stock symbol for AT&T has been "T," and this continues, notwithstanding divestiture. With all of the competition in the long distance market and everything else that is going on in the telecommunications field in the United States, I would still stick with Mr. "T" for long-term capital investment.

IBM

International Business Machines Corporation (IBM) (NYSE) is not strictly a telecommunications company, but as the world's largest manufacturer of computers and information processing equipment and systems, it's becoming very much involved in telecommunications and information processing. I'm sure that IBM will be moving further into telecommunications as the Information Age goes on and particularly as data base programming becomes an integral part of almost every business, not only in the United States but throughout the world. This has been confirmed by IBM's recent acquisition of Rolm Corporation, one of the largest companies in the telephone equipment business in the world.

In addition, IBM is one of the partners, with Aetna Life and Casualty Company, in a partnership called Satellite Business Systems. This company is offering voice and data transmission services to large corporations and small business, but it has found the going somewhat difficult because of competing services. It has one of the most reliable and least costly long distance services for equal access purposes. Whether or not SBS is successful (I happen to believe that it ultimately will have a profitable piece of the market with or without Comsat, which pulled out of the SBS partnership for financial reasons), I'm sure that IBM will benefit most from the growing utilization of personal computers in offices and in residences, in linkage with other telecommunications technologies for transmission of information and to operate various systems in and outside of offices and residences.

IBM had gross income of $40.2 billion in 1983, and had earnings per share of $9.04, which is pretty good, and 1984 probably was even more profitable (over $30 billion of gross income as of September 30, 1984). Profitability can only increase as the Information Age matures.

IBM is ably managed by John R. Opel, chairman of the board and chief executive officer, and John F. Akers, its president. IBM's common stock sells for over $100 a share on the New York Stock Exchange but is worth every dollar paid. The question is whether or not it will hit $250 per share before the decade is over. I think it will, if it isn't split first, in which case it will have an equivalent market value.

DOW JONES AND COMPANY

Last but not least on this list is a company that presently is involved in telecommunications in only a small way, through a minority interest in a cable television system operating company. It is best known and derives most of its revenues from publishing the *Wall Street Journal,* probably the best newspaper anywhere, *Barron's,* and books on business subjects. Its product quality is without peer. Dow Jones will be using its expertise to fulfill the growing demand for news and information delivered by telecommunications technology. Its Information Services Group, which is very much involved in electronic publishing and now is in the black, will be producing significant profits for the company and its shareholders in the not-too-distant future.

Under the wise and dynamic leadership of Warren Phillips, its chairman and chief executive officer, and Ray Shaw, its president and chief operating officer, Dow Jones should have sizable revenues and significant earnings for a great many years to come.

Honorable Mentions

A fairly low-priced stock and some good television capabilities lead me to believe that Reeves Communications Corporation (OTC) may be a pretty good investment these days, although it will have to increase its international sales to be a larger factor in the programming field. Reeves has the capability to be prominently involved in producing programming ("software") for television and for the various new technologies, and its shares should do much better if it does so successfully.

Some of the companies formerly involved only in newspaper publishing have now entered into telecommunications in one way or another. Both the Times Mirror Corporation (NYSE), based in Los Angeles, and the New York Times Company (ASE) are looking toward producing earnings from owning and operating broadcasting stations and cable television systems. Times Mirror, most famous for publishing the *Los Angeles Times,* also publishes *Newsday* on Long Island, New York, the *Denver Post,* the *Dallas Times Herald,* and the *Hartford* (Connecticut) *Courant.* Times Mirror also owns seven television stations, and cable systems now serving almost 1 million basic subscribers. Times Mirror had a two-for-one stock split early in 1984, and I look for it to grow substantially in the telecommunications field.

I feel the same way about the New York Times Company, although perhaps with a little more hesitation. The Times Company now owns three television stations, WQXR-AM and FM (the great but not very profitable classical music radio stations in New York), and a large cable television system in the southern part of New Jersey, which it purchased a few years ago for what seemed an extraordinary amount of money from Irving Kahn, the brilliant cable industry pioneer. The Times Company had over $1 billion in revenues in 1983, with earnings per common share of $2.02, and I expect that it will continue on its upward trend for many years.

Because of their publishing capacities, and particularly the talent of the excellent writers and editors on their staffs, both of these companies have the ability to provide a good deal of the material to be carried by the new telecommunications technology for informational purposes, a business area that ultimately should be quite profitable.

Fortunately, the news isn't all bad for Western Union Corporation (NYSE), the holding company for Western Union Telegraph Company, a regulated common carrier that provides vast telecommunications services throughout the world, including a new one for electronic mail, named EasyLink. Western Union isn't as good a company as some of the competitors already named as the best, but profitable business may be expected as it markets its various services.

If you like good electronic equipment, such as a Professional Walkman, a good digital AM/FM and shortwave radio, or the Trinitron color television set, you probably own as many SONY Corporation (NYSE) products as I do.

SONY makes a very large variety of consumer electronic equipment in addition to marvelous professional products used in the communications business, and it clearly is one of the world's most outstanding companies in the field. However, the SONY Corporation has made a few serious errors in judgment in recent years in selection of technology to manufacture and market, particularly including the Beta form of videotape recorder. Its net sales have dropped off from a 1981 high of $4.8 billion, but sales and earnings rebounded by the end of its 1984 fiscal year. However, with the growth of use of video equipment, including that used by professionals, SONY, led by the dynamic Akio Morita as chairman, is an excellent long-term stock to have in your portfolio. It has its American Depositary Receipts (each of which represents one SONY Corporation common share) traded here.

Well, that's it—the list of Hamburg's best buys, and even some honorable mentions, in the telecommunications field. I can't say for sure that any of these are bargains (as I've said, I'm not one for penny stocks), or that it is certain they will all increase in value, especially in a volatile stock market. In my opinion, however, your chances for appreciation are better in telecommunications than in almost any other industry in the world. Have fun, and let me know how you've made out—at least by 1990.

Now that we've satisfied some of the interests of the *passive* investors in certain public corporations, let's see what can be done for those of you who would like to be *active* in some aspect of the telecommunications business. If you want to know how to be a passive investor in a privately held entity, your interest can be pursued by looking beyond to Chapter 4, where we consider how others can set up their own telecommunications companies with some of your money.

To Really Make a Lot of Money, You Have to Be an Owner

If you want to be an active participant in some aspect of the telecommunications business and make *a lot of money* doing it, you *must* be an owner. It's also the best way I know of to avoid sexual prejudice, so I advise women not only to set their sights on getting into the telecommunications business, but to think about being the owner of or a managing partner in that business. That way you can control your own destiny and decide both how the firm should be managed and how much money you earn. You never have to be beholden to anyone if you are the boss, and you don't have to be concerned about sexual advances from the president of the company if *you* are the president of the company.

You can work for other people all of your business life, and the probabilities are that (with some exceptions, as we will see later) you never will make as much money as you could being a part of the ownership team. This is true in virtually every aspect of the telecommunications field. However, how much ownership you can keep in any enterprise these days depends on many factors, the most important of which is how much capital you have to begin with.

Clearly, if you have only $10,000, you are not going to have very much equity in a new telephone company. But with that amount of capital and a bank loan to buy merchandise, you might be able to open a store in your town to sell telephones and related equipment. You might also be able to set up a radio station in a small but growing town with approximately that amount of your own capital and still be able to keep almost 100 percent of the equity in such an enterprise. You could be your own boss (you might even get your spouse involved) and do something every day that you enjoy doing. You might also see the town grow, and as your station sells a lot of advertising, enjoy the benefits of significant operating income from which you could take out a good salary (and perhaps some extras, such as a new car leased from the local Lincoln dealer with whom you trade advertising time). You could even sell your station some-day for a sizable multiple of its gross revenues of cash flow and thereby make a large capital gain.

Similarly, you probably would not be able to buy a cable television system with only $50,000 of your own capital. But with that amount you could obtain some limited part-ners who might realize tax shelter benefits and profits by investing their money with you in such an enterprise. You could become a leader of a group of cable system owners, take a meaningful salary, and continue to acquire other cable systems until you have a group of systems, so that you are then a multiple system operator (MSO). From that you either could continue to enjoy the benefits of being the manager of a multiple system or sell out at a high multiple of the number of subscribers the systems have and reap your fortune.

You may hate the kind of programming you (and perhaps your children) are forced to see on commercial television or cable or the materials you are almost forced to buy or rent on videocassettes. You may think that you can produce better, or at least more interesting, programming for televi-sion or cable networks or for videocassette distribution. You only have, or at least you know where to get, $25,000 to start your own production company. You're talented and full of ideas, and you know it, but you haven't the faintest idea where to begin to become an active entrepreneur in the programming field. Well, it probably will make you feel a little more optimistic when I tell you that it has been done by many, many people, including some of my clients over the years, *and it can still be done today* if you have the talent, the patience, and the fortitude.

Not too many women have organized such ventures yet, but the time is ripe for women to take hold, especially in ownership, programming, and high-level management. I believe that we would have a lot better broadcasting in this country if more women got more involved in the industry.

How much initial capital you may require for any enterprise depends, of course, on *what* that enterprise is and *where* it's going to be conducted. You could not start a new radio station, even if a frequency were available for one, in Minneapolis, Minnesota (with a population of 2,178,000, and where retail business totals $14 billion a year), or San Diego, California (with a 1,997,000 population and $11.4 billion of retail business), for the kind of money that you could in Brattleboro, Vermont, or in Puyallup, Washington, where there are populations of less than 100,000 people and business totaling maybe $100 million a year. In the same vein, it would be out of the question these days to commence operating a paging company anywhere with less than $250,000 capital, because of the high expense of equipment and sales and promotion required for such an enterprise. But you could start with your $10,000 and hopefully set up such an enterprise if you got additional backing from investors.

With the proliferation of various kinds of telecommunications properties already in place, the easiest and probably best way of becoming an active participant in the broadcasting business—and of increasing your asset value—is to acquire an already existing property. Then you use your (and others') abilities and capital to build it up and either acquire more until you have a sizable empire or sell off the property or properties for appreciable profit.

Or you may want to invest anywhere from $1000 to

$500,000, still have a little say in the operation of the business, but not have to worry about managing on an everyday basis. You are the perfect person for the activist manager to meet, and hopefully you can get together. In Chapter 4 we'll see how that kind of financing can be arranged. However, keep in mind that if you are not a United States citizen and want to invest directly in broadcasting, you cannot own more than 20 percent of the licensee entity.

Consider the Opportunities

No one can tell you *all* of the various forms of arrangements you might want to consider to get into some aspect of the telecommunications business. But I certainly can tell you about some important ones that could lead to increasing your net worth, if not to riches, provided that you do things in the right way and have the right stuff to see your venture through.

Here are just a few opportunities you may want to investigate.

ADVISING REAL ESTATE DEVELOPERS

You may want to consider setting up what is called a shared-tenant communications system consulting service, to assist building owners and real estate developers in comparing and deciding on computer networks, telephones, and other telecommunications systems. In doing this, you might even become involved in the original design of a building or perhaps an entire project. If you know enough about what has to be done, you can help develop computerized facilities management and teleconferencing in all kinds of real estate locally, regionally, or even nationally.

TELECONFERENCING

Teleconferencing is becoming increasingly a part of large business operations and government operations. Those who know how such facilities can be set up—using video and audio equipment together with telephone lines, satellites and earth stations, and perhaps microwave transmission equipment—can get in on the ground floor of a burgeoning business.

BECOMING A
TELECOMMUNICATIONS SYSTEM OWNER

You can't forget the three largest forms of telecommunications today, aside from telephones. Radio, television, and cable television are very exciting businesses, with enormous growth potential for investment and profitability. They also offer a chance for you to make a meaningful contribution to our society as well. But before you can become an owner of such facilities you must learn something about the business and the technology. Probably the best way to learn, if you aren't in the business already, is to take a job in a station or a cable system. Many colleges and universities offer communications courses, but you will find that while they are useful, they don't give you very much *practical* knowledge. You still have to work somewhere to find out what it's really all about. But you can get a lot of useful information short of hands-on experience from a diversity of sources.

You Have to Learn About the Business

The most vital step for anyone who is interested in becoming active in the telecommunications field is first to learn more about what it is, especially from a technological point of view. This applies to almost every aspect of the industry. You should not invest your own or anyone else's money in a venture in which you have little idea of the technology involved.

EDUCATION RESOURCES

To gain that knowledge, I suggest that you attend some of the conferences on new telecommunications business opportunities given by such organizations as the following:

- *The Telecommunications Conference Center,* a division of The Energy Bureau, Inc., 41 East 42nd Street, New York, NY 10017; (212) 687-3178
- *TeleStrategies,* 6842 Elm Street, McLean, VA 22101; (703) 734-7050
- *The Software Institute of America,* 8 Windsor Street, Andover, MA; (617) 470-3880
- *Phillips Publishing, Inc.,* Suite 1200N, 7315 Wisconsin Avenue, Bethesda, MD 20814
- *The Center for Advanced Professional Education,* 1820 East Garry Street, Suite 110, Santa Ana, CA 92705
- *The Institute for Advanced Technology,* 6003 Executive Boulevard, Rockville, MD 20852; (800) 638-6590

If you call or write these organizations, they will describe to you fully the courses and materials that can educate you about the engineering aspects of new possible enterprises and/or give you some business training as well. Some of these organizations also offer excellent publications on business aspects of telecommunications. You can expect to pay rather high prices for publications like TeleStrategies' *1984 Bypass Market Analysis and Demand Forecast* ($1,495) or the *1984 Satellite Transponder Supply/Demand Analysis and Directory* ($995) or for even a one-year subscription to *DataCable News* ($287) or Phillips Publishing's *1984 Interconnect Survey* ($127) or *Finding Telecommunications Information* ($595)—but they probably are worth it.

You might also want to read and subscribe to such excellent periodicals as *Communication News,* a Harcourt Brace Jovanovich monthly publication, which really covers the communications industry in a very comprehensive way, or a fairly new well-done biweekly publication, *Communications Week,* a CMP Publications, Inc. product.

These seminars and publications will give you a much better idea of the workings of such telecommunications subjects as:

- *Voice communications*—including how the telephone works, call processing, microwave, satellites, switching, and various other telephone services;
- *Data communications*—the different types of facilities for transmitting data over communications networks;
- *Office automation and PBXs*—electronic office products and how they work, including telephone exchanges, facsimile equipment, voice-data exchanges, and electronic mail facilities;
- *Advanced data network services*—including packet switching, packet carriers, and the economics of setting up a network;
- *Local area computer networks*—the media of transmission, including baseband, broadband, fiber optics, and other transmission considerations, as well as what equipment is offered by manufacturers;
- *Satellite communications*—the basic technology and its applications, including voice trunking, high-speed data, facsimile, cable television program distribution, and corporate networks;
- *Telephone bypass*—the conventional systems, private microwave, mobile communications (including cellular radio-telephone and digital paging), fiber optics, and two-way cable television.

Addresses and prices of a variety of helpful publications are listed in Appendix A.

The Business Plan

On the assumption that you've learned all there is to know about the area of telecommunications that you want to be in, and you are still determined to become an owner, then the next step is to have a business plan. The largest companies have plans and budgets both for their own forecasting and for their investment bankers, investors, and lenders, and you will need such a plan, too. I have seen many enterprises fail—even with marvelous ideas behind them, zealous personnel, good equipment, and a lot of money—because the entrepreneur never bothered to develop a business plan. To the entrepreneur's ultimate detriment, no one forced him or her to do it—not the investors who gave the developer the venture capital, not the bankers who loaned the enterprise long-term and revolving credit money.

Regardless of whether you want to set up a telephone equipment company or a local paging operation, buy radio or television stations or cable systems, or want to set up your own programming production company, *you must have a plan*. For this you should be able to say what you estimate it will cost the enterprise to enter into the business, and indicate:

• What personnel, space, and equipment will be required;
• What kind of gross and net revenues will be produced;
• What exactly you estimate the operational expenses will be;
• What amount of cash flow and profitability may result;
• What kind of depreciation can be deducted from gross profits for capital equipment;
• What kind of payments can be made for financing the enterprise, including amortization of principal and periodic payments of interest;
• And, yes, what amount of taxes will have to be paid.

To aid yourself and others who want to help you, you should be able to make projections for *at least* five years ahead. (I recommend that clients do it for at least seven years down the road.) Sure, you are going to be making some rash guesses about the market, your sales abilities, or even the overall economy, but if you are knowledgeable, you'd be surprised how accurate you can be.

Many people have come into my office with ideas for businesses, not just in the communications field, but in many other areas, all steamed up and raring to go. They want to have a corporation established so they can open a bank account, put their (and often relatives' or friends') money in, and start a business. When I ask them what their plan is and where they expect to be two years from now, they look blank. When they ask me to help them arrange some bank financing for their enterprise, I tell them that they can't do it without a plan, that they must have one before they start taking in other people's money or even spending their own. Very often they tell me, "Oh, drawing up a plan is no problem. I'll have one to you three days from now."

Sometimes three days stretches into three months. Quite often, it becomes three years, and the enterprise never gets off the ground. They simply have no idea of where they are going, and often sitting down to sketch out such a plan makes the person or group realize they may very well be pouring money down the drain if they start the enterprise they intended to set up.

I don't mean to discourage anyone, and I know that if nothing is ventured nothing can be gained. I'm simply saying that if you're going to become involved in a business, you have to do things in a businesslike way, both for yourself and others. That businesslike way, in my estimation, requires a good business plan. If you can't do it yourself (and I don't think you should try it if you don't have the professional qualifications, either with business school, corporate finance, or accounting training), you must work with a qualified accountant and, hopefully, an experienced attorney.

Once you have them working with you, you can:

• Write up what the market for your product or service is.

• Indicate what you want the business to be earning initially and over a period of time, including gross revenues and expenses as well as a ratio of operating income to net revenues that will give you the kind of cash flow you need and, hopefully, a good internal rate of return on the capital invested.

• State what kind and the number of personnel it will take to operate the enterprise on a per annum basis.

• Set out how you expect the business to progress (perhaps 10 percent growth in gross revenues per year?).

• Carefully chart out what kind of capitalization you think will be required initially and as you continue your growth.

You can (and should) then prepare "projected pro forma financial statements" for the enterprise to be included in your plan.

Have this plan of yours typed up and then review it over and over again with your accountant, lawyer, anyone else you know who may be experienced in the field (and who you trust not to be a competitor), or with an investment adviser or someone else whose business judgment you respect. Make as many changes as you think are necessary, and put the plan into a neat folder so that you can then show it to others.

There's a new personal computer program for preparing a business plan which I rate very highly. I have found it to be an invaluable tool, especially in preparing pro forma projected financial statements. It is the *Lotus 1-2-3* program. Combined with a good word processing program such as *Multimate,* it really helps you to write out your plan and envisage not only the words but the figures with great facility. Before you know it, you'll be thinking like a top pro business manager. I am sure you will find everything much more organized and clearer for you as you project your thoughts into the computer. (Obviously, no program is any good without the computer to run it; I find the IBM PC the best of the lot for normal, everyday usage, although for portability you might want to consider the Compaq, which is fully IBM-compatible and has good built-in graphics capability. This book is being prepared on both of these computers, and I use both in my office and at home for my clients' and my own work.)

In Appendix C I've given you an example of a business plan's financial projections in a company established to purchase radio stations, so that you can see how this is done. The business plan itself can take many shapes, so you will certainly have to construct your own, depending on the aspect of the business you want to carry on. That's where the experienced accountant and lawyer will really be helpful, if you haven't already learned how to do this yourself in a business school. Books exist on how to write a business plan, and you may want to consult one of them. A number are listed in Appendix A.

Financing Your Enterprise

It won't do you any good to go through all of the effort of learning about the telecommunications business if you don't have the financing available or aren't prepared to go out and get it. It's necessary to start the business and carry it on. Because this is such an important element of getting into the telecommunications industry, look ahead to Chapter 4 for some thoughts on capitalization and financing.

Getting Started—Establishing Something New or Purchasing an Existing Property?

There is no point in my trying to tell you the intricacies of setting up *all* of the various forms of enterprise that you may consider in the telecommunications industry, especially when it comes to all of the new technologies. So many things are happening so rapidly that before I could tell you about all of them, they would probably change.

I would like to tell you about three specific areas of telecommunications that I know something about and explain how you can get into ownership in any one of these, either by establishing a brand-new facility or by purchasing one or more that already have been established by others. Radio, commercial television, and cable television have a continuing potential for significant growth. Millions of dollars are being made in each. I *heartily* recommend getting into ownership in any one (or even all three) of these categories.

So that you can see how you could enter any of these businesses, I will give you some examples of others who have done it and explain to you some of the tricks of the trade. From these examples you will see that different amounts of original equity capital were involved, together with a variety of forms of financing. By observing what worked for others you can decide for yourself just what amount of money you should get together in order to become an owner yourself. Unfortunately, none of these examples involve women entrepreneurs (except one who was and continues to be an invaluable coentrepreneur with her husband). However, as I've said, I know that what men have done before, women can do equally well, if not better; there should be *no* hesitancy on the part of women to try to become owners.

If you are a novice in the field, I strongly advise you to read everything that you can about the industry, including the magazines and texts listed here. The magazines in the field that I like best include *Broadcasting, Television Digest, Variety, Electronic Media, Radio and Records, Radio Only, Television/Radio Age, CableVision, Radio News, E-ITV News,* and some of the publications issued by the National Association of Broadcasters, such as *RadioActive,* the *Accounting Manual for Radio Stations, Buying or Building a Broadcast Station* (formerly *Buying and Selling Radio Stations*), an excellent text by former NAB Senior Vice-President and General Counsel Erwin Krasnow.

You can get started by teaming up with someone who has experience in the field, or start by getting a job in an existing organization in sales, programming, or management, where you learn as much as you can before you go about becoming an entrepreneur.

Let's start with the most popular, least costly, and perhaps easiest area of telecommunications first—radio.

Getting into Radio Ownership

I like the radio business a lot, not just because I've always imagined myself a great sports announcer (nearly everyone who knows me is aware that this is my great fantasy), but because I believe that radio is the steadiest of all the telecommunications businesses. *Everyone* listens to the radio at some point in the day—99 percent of American homes have radios—since you can do a lot of other things while you listen to the radio. You can't watch television in the car, and people spend a great deal of time in their automobiles. You can't take the cable to the beach or to the countryside when you go camping, but you sure can take a radio. Out of approximately 470 million radio sets in the United States, it is estimated that 123 million are used outside of the home.

The Radio Advertising Bureau estimates that by the end of 1983 there were more than 9000 commercial radio stations operating in the United States, about 5000 of which were commercial AM stations and 4000 of which were commercial FM stations. There were also over 1100 noncommercial FM stations. Recent statistics available from the U.S. Department of Commerce indicate that radio billings (meaning advertising time sold either on a national, regional, or local basis on all commercial radio stations) totaled around $4.5 billion, with over $3 billion of these sales coming from ads placed by local advertising agencies and through direct local sales. That seems like a pretty good business to me, and one that you might very well want to get into—and can—without having or spending very much of your own money.

Many people I've met in the last five years tell me that they long have wanted to own and operate their own radio station. I think I know why. Either everyone believes that he or she can be a great radio personality and thinks the best way to get on the air is to become an owner, or they are certain they know how to program a radio station much better than "those morons" who are running the station (or stations) they now listen to. Well, maybe they're right, but there seem to be a lot of people making a lot of money in the radio business, so they must be doing something right. Fortunately, however, there are also a lot of people doing a lot of things wrong in the radio business, and that's where opportunity knocks for the venturesome.

There is a very enterprising and diligent man named James H. Duncan, Jr., who lives in Kalamazoo, Michigan. He obviously has a great interest in the American radio business. He has for the past few years done an excellent job gathering data concerning all of the radio stations in the United States in a rather fat mimeographed text called *American Radio*, which he sells only by mail subscription for $58 and which appears semiannually.

Recently, Jim came out with the first edition of his *Duncan's Radio Market Guide* (selling for $110), and it really is something that anyone interested in getting involved with radio ownership in America must read. It contains histories of all the AM and FM radio stations in the United States and the condition of each radio market, including total radio revenues by market in 1978, 1983, and as projected for 1988. It also lists the fifty markets with the highest revenues in those years and many other facts that anyone interested in the radio business should have at his or her fingertips. Duncan's *American Radio* also lists data on the industry and stations, including all of the stations, their formats, their Arbitron ratings over the last five spring and fall rating books, and much more. You can order these publications by writing to James H. Duncan, Jr., Duncan Media Enterprises, Box 2966, Kalamazoo, MI 49003, or by calling him at (616) 342-1356.

The *Broadcasting/Cablecasting Yearbook*, which is published annually by *Broadcasting* magazine, also offers a wealth of knowledge about the radio industry, and contains comprehensive materials on television stations, networks, and cable television systems. You can find in it much of the information needed to consider entering the field, including a listing of broadcast engineers (who can be very helpful when you consider buying a broadcasting property), broadcasting brokers, and communications lawyers. *Television Digest*, which is also a trade news weekly, publishes a similar annual directory. Both of them are issued from Washington, D.C.

A list of the foremost publications in the telecommunications field, with addresses and subscription prices, is set forth in Appendix A in the back of this book.

Starting a New Station

How do you find a radio station to buy these days? It really is a difficult problem, because many of those already established have become so valuable, particularly those that are in major market areas. Most owners are reluctant to part with these stations. So to begin your entrepreneurship, you may want to consider trying to set up a brand-new station in a small town that has none or few, or even in an area that has a few stations but not one that is really outstanding. This can still be done, although it is a very hard task, and you should be prepared to suffer a lot with such an enterprise before you see any real signs of success. Even so, starting something new can be fun, especially if you're young and courageous and you think you have the personality to bring it about.

The FCC has been helpful to newcomers. It has adopted some new rules that expedite processing of applications for new stations, particularly FM stations using certain frequencies. The National Association of Broadcasters, located at 1771 N Street N.W., Washington, D.C. 20036, has also started some programs to help new entrants into the field, especially minorities. These new "assignments" of FM frequencies are part of what is known as the "80-90 proceeding" in process at the FCC, and hopefully this will be fully concluded shortly so that a lot more stations can be started by 1986.

If you are interested, then by all means try for it! You will find, as have many others before you, that you can get a lot of cooperation from a lot of people and organizations, including banks, if you are organized and businesslike. A couple I know, Jerry and Sasha Gillman, did just that a few years ago in the town of Woodstock in upstate New York.

Jerry Gillman had been in the public relations business in New York City, and he knew something about advertising and marketing but not too much about the operation of a radio station. He read a lot, he talked to a lot of people, and he saved his money. With the help of a lawyer, an accountant, and a friendly banker, the couple applied for a construction permit to build the first FM radio station the town ever had. They realized that the town was growing. They liked living in Woodstock, in what used to be their vacation house. It became their primary residence when they finally received the permit (it did take a long time). They first had the station's antenna constructed, then put their transmitter into a shack at the antenna site, and installed their broadcasting equipment, studios, and offices in an old building in the center of town.

Jerry thought the people in the area would like to hear good music and news and information, so he went on the air as a commentator and sometime disk jockey. His charming wife suddenly became the treasurer and the person in charge of all of the business. They did everything they could to promote the station everywhere in the area. In the first year they managed to sell some advertising time on their station and almost broke even. In the second and third years they did progressively better, and they became very much a part of their community.

Jerry and Sasha are now contemplating building another station in a nearby area, and I think they will do very well if they are successful in obtaining their construction permit. If you ask them whether they've had a good time in the radio business, they will go on for an hour about the pleasure having their own broadcasting business has given them. If they keep on building new stations and operating so successfully, they might even end up having really solid capital, something they didn't envision when they first made the big decision to enter into radio ownership.

You can do the same thing. But the first thing you must decide upon is location. You can't simply build a station anywhere. FCC regulations exist to enforce an orderly use of the crowded radio spectrum. Moreover, you certainly will want to determine what prospects there are to guarantee the economic success of such an enterprise. Take a good look around if it isn't a town that you have lived in or worked in all of your life. Find out as much as you can about the demographics of the town. How many men, women, and children are there? What age and income categories do they fit into? What are the growth prospects for industry and commerce in general, and by how much is the population projected to grow in the next ten years? Determine the kind of money that may be spent in advertising, what growth prospects there are, and what kind of people live there. If possible, find out what their listening or viewing habits are and where there may be a void that could be filled by a new station.

Assuming that you have made the decision to set up your own station and decided where you will locate, you will want to take a number of steps

First of all, contact a communications lawyer. You need

experienced counsel to help you prepare the necessary documents to apply to the Federal Communications Commission for the construction permit required to build the station.

Find a qualified engineer to consult. Your lawyer may be able to help you. You want a knowledgeable professional who can locate the best available location in or at least near the town to put up your transmitter and tower(s). If you are building an AM station, you will also need land for a ground system as well. This will take some time, and you will want to have everything planned out and in place once the permit is awarded so you can begin the construction. If your antenna location is near an airport, you may have to apply to the Federal Aviation Administration for permission to construct under certain conditions.

Have an accountant help you set up whatever financial statements are required. As the lawyer and the engineer help you prepare your application for the construction permit, have your accountant organize the fiscal records and affairs of the entity that will be the applicant.

Discuss the needs and interests of the community you hope to serve. In preparing your application, and maybe even as part of the entity-organizing process, you will want to meet with as many leaders of the community as you can. You might even want to ask a few of them to serve on your board of directors if you are using a corporation as the potential licensee. It certainly can't hurt to have a leading banker join you.

Upon receiving approval, construct your station. It will take some time (and ideally there will be no competition for the same permit and frequency you wish to use), but I assume that you will be successful in obtaining the construction permit. Then all you have to do is construct your station as you said you would within the time period required by the FCC.

Apply for your permanent license to go on the air. After construction is completed, and you have carried out the necessary tests, you will have to apply, again to the FCC, for the permanent license to go on the air. Fortunately, these radio station licenses are for a longer term than they used to be (seven years rather than three), and the renewal process has been made much easier for radio licensees. That doesn't mean you don't have to abide by established rules and regulations, but these are a lot less cumbersome than they used to be, and your life as a licensee has been made much easier through the continuing deregulation that has taken place under the Reagan Administration.

If the station or stations that you want to build and operate from scratch is or are in an area with very obvious potential, you may find six or seven persons or groups competing with you for a construction permit in the same community. Be prepared to spend time and money in trying to win such a competition, and decide if the effort and expense are going to be worth it if you are successful. Sometimes such proceedings, which may involve comparative hearings at the FCC, cost very little (under $5000) if the matter is not an intricate or lengthy one, but they might end up costing $500,000 if the competition goes on for three or four years, as some have.

As an example of this, the FCC assigned a new FM channel to the town of East Hampton, a very affluent, mostly summertime community on the southeastern part of Long Island, about two and a half hours from New York City. The nearest FM station operating in the area is in the town of Southampton, about fifteen miles west, also essentially a summer community, although now with a growing permanent population. The Southampton station can be heard quite clearly in East Hampton, as can two stations located in the old whaling village of Sag Harbor, a few miles away, and a number of stations in Connecticut, across Long Island Sound. Although none of the radio stations in the area are very profitable, no less than seven persons or groups have filed applications to construct an FM station in East Hampton. To compound the problem, the town fathers in East Hampton have indicated their resistance to setting up an FM tower in any part of their lovely community, and none of the applicants have been able to determine just where they could place their tower if they were to be awarded the frequency.

Since the FCC's rules require a comparative hearing if there is more than one applicant, in all likelihood such hearings will have to be held. Because of a rule concerning the desirability for media diversity, any applicant who owns other stations, or even other media, will not be as qualified as one without such ownership background, but ownership by women and other minorities is encouraged. The matter will have to be referred to an administrative law judge in the FCC. All of the applicants will be allowed to examine the applications and qualifications of the others, and a final determination will be made sooner or later. This will probably take at least two years, if the applicants are lucky. And there is no certainty that the successful applicant will be able to operate profitably a station in a community where most residents live for only two or three months of the year.

As you can imagine, such a start-up operation is not an easy matter. It can only be successful if you are prepared to own and operate the station or stations yourself. Such owner-operators can be right there all the time to see what is being done, to manage sales of advertising time (which is the essence of the business), and to watch over expenses. The bottom line, or broadcast operating income, depends on how good a manager you are.

Buying an Existing Property

Let's suppose that you don't want to start a new station from scratch but still feel that getting into the radio business is something you want to do. Well, you can do either of two things. You can buy a station (or perhaps a combination of AM and FM stations) that someone else started and is operating but that is not doing so well—has very little cash flow or is even losing money—in the belief that you and/or your associates have the ability to turn it or them around. Or you can play it a little safer by buying a good station or a combination in a good (and hopefully growing) market with good ratings, good sales, and good cash flow in the belief that you can increase operating income because of growth in the area or through your (or your associates') ideas and zeal in improving personnel, programming, promotion, and advertising revenue.

I happen to prefer the latter approach to buying existing properties. I think in the long run it proves safer and more profitable, especially if you don't have, or don't want to spend, a great deal of money (your own or other people's) in starting up the business. Again, you may want to do careful research to determine what kind of radio station you will want to own and what the values are.

Just what kind of research can you do? About the same as you would if you were going to set up a brand-new station. But your job will be made a little easier, since you are not starting from scratch. You should obtain materials about the community from the local Chamber of Commerce or any development corporation, local or state, that is trying to encourage growth in the area. Local lawyers and bankers, as well as the builders, restaurant owners, and other entrepreneurs, are good people to talk to to ascertain how stable the community is, what the nature of the populace is, and what kind of incomes they have. Of course, it is worthwhile to listen to other stations in the community in addition to the one you may acquire to see how you might fit in and make your station better and more profitable.

An outstanding entrepreneur by the name of Paul Kagan has established quite an enterprise in the telecommunications field in Carmel, California. He sets up seminars on the industry that are given periodically in a number of cities; he also markets texts on the industry. One seminar is on radio station values. If you can't attend a scheduled seminar, you might want to purchase some of the audiocassettes generally offered with the written materials that are handed out at each seminar. I have never attended a Kagan seminar, but I've been told that they are stimulating and informative. PK Services Corporation is at 26386 Carmel Rancho Lane, Carmel, California 92923; (408) 624-1536.

If you are interested in purchasing radio stations in the United States, I recommend that you look around for the best possible stations in a growth area. This need not necessarily be in what is called the Sun Belt, but it doesn't hurt to try to acquire a property in that area of great potential. Sometimes a city or town in the Northeast or Northwest can be just as attractive because of an increase in new manufacturing or high tech business, or even renovation of a formerly blighted downtown area.

Assuming that through great diligence or through a good radio broker you have found the property that you think you want to acquire, there are several important followup steps to take. (Brokers, by the way, almost always represent the *seller,* from whom they will be receiving their commission, so carefully check all of the facts given to you by the broker—just as you would if the information were to be given to you by the seller.)

Ask for audited financial statements. Get them for the last three years of operation of the station or stations. If you can't get *audited* statements, then you should obtain some formal representation from the owner when the financial information is sent to you that the statements are true and complete. These representations should then be written into your contract when you make the purchase. You should, at the outset, also get up-to-date figures on operations since the end of the last full year, including an aging of the accounts receivable as current as possible. This aging will show what part of the station's accounts receivable are current or haven't been paid in thirty, sixty, ninety, or over ninety days from the date of billing.

Have the financial records carefully reviewed by a knowledgeable accountant and, if possible, obtain permission for your accountant to talk to the seller's business manager and the seller's auditors or accountant so any open questions can be answered. This is best done immediately after you receive the seller's financial statements.

Have a qualified engineer look at all of the station's equipment. This includes transmitters, towers, studios and production facilities, and an examination of the station's "sound" in comparison to the sound of the competition in the market.

Our engineer in our stations in Providence, Rhode Island, is Dana Puopolo, who now runs his own consulting service in Hyannis, Massachusetts. He recently wrote an informative article for the NAB's *RadioActive* entitled "Buying a Station? Don't Forget to Kick the Tires." I think that's virtually what you have to do.

To quote Dana, "Many owners consider the equipment in their radio stations a necessary evil. As a result, many stations have little or no maintenance performed on equipment. The attitude of some people is, 'If it works, leave it alone.' This attitude, while always prevalent in small and medium markets, is now appearing in the larger markets as well—usually in the lower-rated stations. A purchaser of one of these stations would do well to consider a revamping of the station as part of the cost of acquiring it."

There are lots of good questions that can be asked about the technology of operating a radio station, whether it is an AM station or an FM station. Don't forget that a great number of our AM stations are what are known as "daytimers," authorized by the FCC to operate only between sunrise and sunset (under recent rules, a little bit before and a little after), so as not to interfere with other stations throughout the country using the same frequency. Sometimes these stations are permitted to operate with a certain power during the day and with less power at night. Many of these daytimers are quite old, as is their equipment, and it may cost more to operate them than the revenues they can produce. They may also require a good deal of money to upgrade their equipment, and it may cost a fortune for electricity charges to keep the transmitter going at the power with which the station is authorized to transmit its signal.

A good, qualified engineer like Dana Puopolo can be of enormous assistance in helping you make a determination on whether or not to purchase a particular property. He or she can even be helpful in deciding on a purchase price and terms, as you find out about the condition of the equipment and how much more money you may have to invest to make the station operate optimally. A good engineer usually can be found by recommendation from other station owners or lawyers or accountants in the field. Most list themselves in the *Broadcasting Yearbook* and will give references upon request.

Find a dependable communications lawyer. This is essential in any broadcasting acquisition. Look for one who combines a knowledge of communications law with acumen in corporate (and perhaps banking) practice, because you will need all these elements in going ahead with your enterprise. A lot of people try to separate out these functions, depending upon what are called "the FCC lawyers in Washington" for some aspects of the deal and a general corporate lawyer for the rest. Unfortunately, some people interested in purchasing radio stations have only consulted a Washington (or New York) FCC lawyer and have not received the kind of counseling they should have received in an acquisition involving hundreds of thousands, if not millions, of dollars. If you can find someone (or a firm) who combines everything and deal with just one person or firm, you will be much better off. The fees will be less, too. Right now there aren't too many of us around, but I believe that this is changing, and the availability of ideally qualified counsel should be greater in the not-too-distant future.

You should be able to discuss your potential acquisition with your lawyer right from the beginning. Don't try to negotiate a deal and then call him or her in to simply write up or review some agreements and file some papers with the FCC. There probably will be a lot of points on which your lawyer can be very helpful while the deal is being negotiated.

Investigate the real estate questions. You certainly will want to know about whatever real estate the station may own or rent. You should query the matter of title to the land and the building or buildings where the studios, offices, transmitters, and towers are.

Review outstanding station contracts and agreements. You will want to have your lawyer look over all the contracts the station has, including those for music that may be supplied to the station by others and agreements with ASCAP and BMI, organizations representing composers to which you will have to pay royalties for playing their music.

A lot of people don't realize that networks *pay* radio stations to carry their programs, or at least the commercials that the networks sell to national advertisers, but they do. You will certainly want to have a lawyer look over these agreements to make sure that you will receive the same remuneration if you continue with the present network or to make sure that you can get out of the agreement if you don't want to continue.

Union contracts, employment agreements, profit sharing plans, programming agreements are among the other contracts the present owners may have that you will want to have counsel review before you finalize any transaction.

Most important, *watch out for any trade agreements* the seller may have entered into that will have to be honored by a new licensee. Very often, the owner or salespeople in a radio station will "trade" or barter advertising time for a week, a month, or an even longer period of time in exchange for goods or services, such as free meals at local restaurants or a year's rental of a new Cadillac for the general manager. These agreements generally have to be honored by the new owner, and they may mean not only that you can't sell new advertising to those who already have trades, but that everyone else in town will want the same kind of deal, which doesn't produce very much cash for a new owner.

Carefully check the sales situation and the kind of programming that the station is carrying.

Make sure that the kind of cash flow that has previously been produced can at least be obtained under your ownership.

It always helps (if you don't already have such a background) to have someone with radio sales experience take a look at the client list and the demographics of the ratings of the station. If you don't have this kind of experience, it also helps a lot to have a consultant who knows about programming actually go to the town the station is in and listen to it. It may prove useful to have some tape recordings made of a full day of the programming of your potential station, as well as that of the competition. Get a report on your consultant's thoughts about the present adequacy of the station and its potential.

THE PURCHASE PRICE—HOW MUCH IS IT WORTH?

Assuming you have done all that is necessary to fully investigate the property you are interested in, and you have decided that you want to buy the station or stations that you have been examining, then comes the big question: *How much should you pay?* Well, my philosophy is to use as little equity as possible, and employ bank loans to buy stations with good cash flow to service debt—and still take something out. Ideally you should pay no more than approximately *eight and one-half times* what the accountants call "broadcasting operating income," which is gross revenues from advertising time sales, less commissions paid to advertising agencies (resulting in net revenues), less operating costs for selling, programming, technical, and general and administrative expenses of the station.

Broadcasting operating income is the "cash flow" that everyone wants. This is what allows you to pay interest on and amortization of long-term and revolving loans, as well as any interest you might agree to pay to investors and any management fees you may want to pay to the parent corporation that you probably should establish to start your enterprise. Broadcast operating income, less depreciation of the station's assets and the amortization of debt, is what results in net profit before payment of income taxes.

Let's assume that a nice FM station in a good market with some good Arbitron ratings in its last rating book is for sale, and you want to buy it. The seller (or a broker) has sent you the last financial statements issued by the seller's accountants. The station has had a good year, and it is throwing off some pretty good cash flow. From the statements sent to you, you extract the following information.

Gross Revenues		$1,950,000
Agency Commissions		(214,500)
Net Revenues		$1,735,000
Technical Expenses	$ 85,500	
Program Expenses	115,500	
Selling Expenses	450,000	
General & Admin. Expenses	549,000	
Total Expenses	$1,200,000	($1,200,000)
Broadcast Operating Income		$ 535,000
Depreciation		(215,000)
Amortization and Interest		(300,000)
Profit Before Taxes		$ 15,000

In talking about the gross revenues and broadcast operating income, you will want to make sure that the figures you are given are for the *last* full fiscal year, *not projections for the year in which you are discussing the acquisition.* The multiple that you are discussing will, in my opinion, only work out if you *use the last full year as a base,* because you are going to make your own projections as to what will occur in the current year and for a number of years to come when you prepare your "projected pro forma financial statements." As observed, preparing such a statement is something every buyer should do for him- or herself, and *must* do for any potential investor or bank from which you want to borrow money. (We'll talk more about this in Chapter 4.)

If there are special circumstances involved in the acquisition (such as the station is in an excellent growth area, has a terrific reputation, and/or is doing remarkable business), you may want to add a little something to the purchase price. You may be willing to pay a *multiple of gross revenues* (anywhere between two and one-half to three and one-half times such revenues), or you might consider paying a little more as a multiple of broadcasting operating income. Some stations have been purchased for up to ten times cash flow by an eager buyer, and quite often proved worth it.

Using the hypothetical figures given above, you could compose a range of various purchase prices to offer. Here's an example of what you should do, although you would have to use the actual data for the *past* year given to you to prepare it precisely.

Actual Gross Income	Actual Cash Flow
$1,950,000	$535,000

Possible Offering Price

2.5×	$4,875,000	8.5×	$4,547,000
3×	5,850,000	9×	4,815,000
3.5×	6,825,000	9.5×	5,082,000
		10×	5,350,000

As you can see when you do this exercise, there are many possibilities for offering a reasonable purchase price that will allow you to carry out appropriate financing and still hope to make a decent profit in the near future. In this situation (assuming the station is a good one and in a growth community), I would try to purchase the station for no more than $4.9 million, which is, as you can see, approximately two and a half times last year's gross revenues and nine times last year's broadcast operating income. Assuming perhaps 10 percent growth per year, you should be able to use the cash flow to finance the purchase (after some equity investment) and still make a profit within two years.

THE ASSET PURCHASE AGREEMENT

Once you have done everything preliminary to actually agreeing on the acquisition, you will want to work with your counsel on the structure of the deal with the seller, either directly or through the broker. Depending upon how you are financing the deal, you may want to consider having the seller help you do some financing by letting him (or it) take back a promissory note from your purchasing entity, hopefully at a rate of interest below that being charged by any financing institution and with a long-term amortization period.

For example, you may agree on a purchase price of $2 million. You may have $100,000 of your capital, and a bank may be willing to lend you $1.4 million on a long-term basis. The difference of $500,000 may be dealt with by your giving the seller a note from the buying entity, which is subordinated to the debt of the bank. (If the assets have to be sold, the bank debt must be paid off first, and then the seller can collect.) A typical note for the $500,000 would provide for payment over a period of five years, with interest *below* the then existing prime lending rate. (A 9 percent note these days is considered quite fair, although under new tax rules, the seller may have to pay taxes at ordinary income rates on imputed interest.)

Many times, the seller wants to get as much as he can for the property and is willing not only to take back a note, but may even let the amortization of the principal begin one or two years after the closing in order to get a higher total purchase price. It never hurts to ask for such terms. More often than not, if the seller likes you and thinks you have a good chance at succeeding in your operation and obtaining even more cash flow than he has, he will accede to such a request.

Similarly, there are tax advantages to paying part of the purchase price in consideration for an agreement by the seller not to compete with you in the same market or within a larger area around that market for a reasonable period of time. Such payments, generally made annually or semiannually, are without interest, are deductible as an expense, and can be spread out over something like five years.

In addition, there may be personnel working for the seller (or perhaps the seller himself, if it is an individual owner-operator) whom you want to keep involved with the station you are acquiring. Having such a person sign a consulting agreement under which you pay him or her a part of the purchase price as a fee, for, say, three years, may allow you to make the payments deductible from your net revenues, and thus decrease the amount of cash that you otherwise would have to pay at the closing or have to borrow for the acquisition.

I always recommend that an acquiring entity not purchase the stock of an existing licensee corporation if it can be avoided, or unless it makes good sense from a tax point of view. Purchase *assets only;* don't acquire the corporation. Then you generally don't have to worry about mistakes made by the seller before the closing or about having to assume liabilities of which you are not aware.

If you can make it a deal involving only the purchase of assets, then you want to have an "asset purchase agreement" prepared. That should incorporate all of the details of the transaction that you have agreed upon with the seller, including how the purchase price will be paid and anything that has to be done before and at the closing. This also will set forth all of the representations and warranties that the parties want (or are willing) to make to each other. That's where an experienced lawyer comes in. Many such agreements that I have seen prepared by lawyers who are not experienced in the business have been badly done and require untold hours of revision, resulting in much higher legal fees than should be necessary.

Because a radio station is a business entity that goes on minute by minute, hopefully with paid advertising running day and night, accounts receivable are constantly being created. Generally, when assets are purchased in a radio station acquisition, the buyer does not purchase the present licensee's accounts receivable, mostly because it involves an extra cash requirement at the closing and the buyer is never sure of being able to collect some other party's receivables. Very often, however, you, as the buyer, want to make sure that there is no problem with existing advertisers and that the seller will not be overzealous in collecting bills due from those advertisers, thus hurting the station's reputation. It's always wise for you, the buyer, to insist on getting a list of accounts receivable from the seller, to be delivered at the closing. For a fixed period of time (I like to say four months) the buyer will collect the receivables, keep the money in a separate account, and turn it over to the seller periodically, preferably once a month, with a full report.

The asset purchase agreement notes that the whole deal is subject to the approval of the FCC; you are responsible for the manner in which applications to the commission will be prepared and filed on behalf of both the seller and the buyer. Again, an experienced lawyer will know what to do about this, but you should be careful to provide for getting your deposit back if there are any problems at the FCC or with any other aspect of the transaction.

THE DEPOSIT

Speaking of deposits, the asset purchase agreement generally provides for a deposit of 10 percent of the total purchase price to be put down on signing of the agreement and then applied to the cash to be paid at the closing. No matter what the amount, I heartily recommend that this deposit be given to one of the party's lawyers or a bank only under a written escrow agreement. If a reputable broker is involved, then the broker may be the escrow agent. But again, the covering agreement should be in writing, and it should specifically provide for what happens to the money if any difficulties occur between contract and closing. These days it is worthwhile to provide that the funds be invested in Treasury bills or money market funds and that the buyer receive the interest at the closing.

THE CLOSING

Assuming all goes well and you receive the necessary approvals of the FCC to become a radio station licensee, then you are going to set up a closing for your acquisition. The lawyers are going to take care of everything for you, including whatever papers are necessary to fulfill your financing and acquire any real estate that may be involved. Probably all that you will be concerned with is having a pen handy to sign a lot of papers and checks. Don't be overwhelmed, as it does take some time and never goes too smoothly. However, you will have to remember that you are taking over a business operation, and this is never easy.

You have to make certain that everything you thought you were buying is still there. You will also have to make deals with personnel who will continue with you or with new people who will be brought on board. Plan ahead before the closing so that if you are going to be changing call letters or formats or policies, everything is in place from the moment the papers and checks are exchanged and your counsel tells you that you are now the licensee. He or she will take care of informing the FCC of the change of ownership and filing the necessary report after.

From then on, it's all yours. For better or worse, you are going to be owning a governmentally licensed radio station (or stations, if you've bought a combination), and you are going to have to abide by the rules and regulations of the FCC, as well as be a responsible member of the community in which your station exists. You are not only going to have to strive to have a profitable operation, but hopefully you will contribute to your community, remembering that you have a powerful instrument of control in your hands.

How to Operate—the Fairness Doctrine

I don't want to tell you how to operate your station, if that's what you've chosen to do, but I believe that broadcasting licensees can make money in their operations and also exercise discretion and fairness in what they put on their stations. In this regard, it may be worthwhile for you to read what has become known as the Fairness Doctrine, which is simply a report put out by the FCC in 1949, after much research, to help radio station licensees and managers determine their obligations under the Communications Act of 1934 (which is still the guiding law for broadcasting in the United States) in editorializing and the treatment of controversial issues of public importance in the community of license. It is a pretty good guide, and I'm sorry to see a lot of people, including some senators and members of Congress, and even the National Association of Broadcasters (NAB), trying to do away with it. In my opinion, it's the best protection broadcasters have against government interference in programming matters; broadcasting owners should try to maintain its principles for the good of the entire industry.

Unfortunately, there has been confusion between the Fairness Doctrine itself and Section 315 of the Communications Act, which essentially provides that equal opportunity must be given by broadcasting stations for political candidates—the so-called "equal time" provision. You can get copies of these from a lawyer, from the NAB, or your senator or member of Congress. They can also be found in the 1949 *FCC Reports* in a law library.

Where the Millions Are

The real money in radio station ownership is in building up a group of stations and either selling them off for appreciable profits or building up the entire operation so that debt is low and cash flow and profits high. I happen to think that the best way to operate is through central fiscal administration. As the multiple-ownership rules, especially for radio station ownership, are abandoned, one group will be able to own and operate as many as a hundred stations at one time effectively and, hopefully, on a highly profitable basis. Computers now allow instantaneous reporting by individual stations to a central controller. Other matters of determination can be made by group administration. This does not mean that local managers lose their importance or that local programming decisions are given up. It simply means that efficiency, which never has been a byword in most radio stations, will be introduced, and that there is a much better chance for making some really big money in the radio business, assuming that you believe in its future— and also assuming the advertisers think so, too.

If you have purchased a station or stations in a small or middle market, "fixed them up," and they are now valued at a much higher amount than the purchase price that you paid for it or them, then you should consider selling off the entity, either to make capital gains or simply to have more cash to purchase other, more profitable stations in a larger market.

Obviously, radio station licenses are not meant to be

commodities to trade around every day, nor will the FCC let you do that. However, there is no reason why good business organizations should not be recognized as such and why profits cannot be made on selling valuable properties if they have been properly managed and the buyer is an acceptable licensee. Such free enterprise has been recognized by the government, and now you can sell even after one year of ownership. This is where many of the millions are being made today in radio ownership. One should never get too attached to any property to preclude a prudent business judgment to sell at a good profit.

Whether you maintain your complete independence as an owner of one or a group of stations and keep your profits for yourself and your investors, or you decide to register and sell stock to the public to obtain additional capital or to add some market value to the shares in the enterprise that you own, you should consider the alternatives as you go along. Much of what you do as you move toward becoming one of those making millions in telecommunications will depend on many factors, the most important of which is your own personal approach to life. We'll talk a little bit more about this in Chapter 4. But all it really comes down to is what you *want* to do, not what you believe you may *have* to do.

The Approach to Television

As I have mentioned, nearly 84 million homes (98 percent of all homes in the continental United States) are equipped with at least one television set. Over 75 million of these homes have color television sets, and 55 percent of all U.S. homes had two or more sets as of January 1984. By the end of 1983, there were almost 1200 operating television stations in this country. Most of these stations are affiliated with some television programming network, but approximately 250 stations are independent of any network.

It has been estimated by the national polling company for television, A. C. Nielsen Company, that there were over 220 million people in the U.S. television households, and that during the 1982-83 television season these households viewed an average of six hours and fifty-five minutes of television per day—the highest average ever recorded.

In 1982, the U.S. Department of Commerce estimated that total commercial broadcasting revenues were $14.9 billion, of which television advertising sales accounted for nearly $11 billion. Over $6 billion of this went to the national TV networks, over $3 billion went to national non-networks, and over $2.5 billion went to local advertising on television stations, according to estimates made by the Television Advertising Bureau.

Don't all those marvelous figures make you want to own a television station? Of course, but not every television station makes money these days and a few have even been known to incur appreciable losses. It's also not so easy to fit within the allocations made for the broadcasting spectrum by the FCC, so you can't simply say "I'm going to build a new TV station," even if you have the experience necessary to run one and the money to build it and to operate for a while.

If you are going to start a station, whether low power (see below), VHF, or UHF (if you can), I would certainly say that the two most essential people are a lawyer and an engineer. Without them you won't be able to do anything. With them, and your good accountant, you may very well be able to start such an enterprise. There's no sense telling you here what all the needs and procedures are, as those professionals will be able to guide you along a path that takes quite a while to complete. Hopefully the end result will be as successful for you as it has been for a few other courageous people of late.

If you are a member of a minority group, you may have better opportunities not only to obtain financing for such an enterprise, but to go through the FCC process at a more rapid pace—maybe even receive greater cooperation in the targeted market from advertisers, talent, production people, and others whose assistance you will need. There are now good available sources of capital, which we'll discuss later.

LOW POWER TELEVISION STATIONS

The Federal Communications Commission has made it a little easier for some people to get into the business, particularly members of minority groups, by allocating certain frequencies for what is known as "low power television" (LPTV) in parts of the country. Unfortunately, a lot of people and established communications companies that were already in the business jumped in and filed for many of these licenses simply to increase their ownership positions, and the FCC is still trying to sort out the whole mess. However, a few LPTV licenses have been granted.

It's still very difficult to get a new station off the ground in any area where you must try to make it profitable quickly. But if you are willing to give it a try for the value that may be there in the end, you may be successful. What I have said about getting into the radio ownership business applies here as well, but there are some special problems you have to consider when it comes to television.

The biggest consideration in putting up a new television station is whether or not the market in which you want to build it can support it. If there are enough stations in the area, and especially if all three of the major networks are affiliated with stations there, you are going to have a hard and costly time buying programming to put on your station, and an even harder job selling enough advertising to make your investment worth your while. This is true even for the low power stations.

Greater Costs, Possibly Higher Returns

The costs are much greater in putting up a new television station and starting its operation than in establishing a radio station, but the rewards can be much greater, too. There is greater coverage, probably fewer stations in the market, certainly more consistent interest in television than in radio, and the possibilities for obtaining advertising dollars are much more significant. The sources of gross revenues may be greater, and national advertising will constitute a much larger percentage of such revenues.

SOURCES OF TELEVISION REVENUES

The sources of television revenues are *national advertising*, which is time sold directly through agencies representing national and regional advertisers; *local advertising*, which is time sold directly to local advertisers or agencies; and *network compensation*, which is the fees paid to the television station by a network for broadcasting its network programs. Sometimes there are other revenues derived from payments for rental of studios and other production facilities.

NETWORK AFFILIATION AND FEES

Probably the least-known fact to those unacquainted with the television business is that the networks pay fees to their affiliated stations to carry their programs. The popularity of the national network with which the affiliated station is associated is an important factor in determining the revenues that the station will derive. If a particular network's programming is more popular than the others', more people will tune in to see the corresponding affiliate's programs and the station's ratings will be increased. This, in turn, increases the rate which the affiliated station charges its advertisers for commercial time during the network program hours. The network program will carry national advertising sold by the network, but generally the station is allocated some time to insert its own local advertising, thus the station receives revenues from both ends.

Stations can change their affiliation as they choose after a certain contractual period, but this isn't done very often. However, lately many stations have found that being independent of a network may be better for them financially, as they are then able to buy or put on their own programs to a much greater extent and sell local time more advantageously. However, there is a lot to be said for a station being affiliated with an ABC, CBS, or NBC, no matter what you think may happen to the networks with the advent of the new technologies like cable. I wouldn't want to rush away from any one of them or dismiss the possibility of such an affiliation if you want to start your own new station.

Buying Existing Stations

I'm always nervous about seeing anyone start something completely new. I would rather pay something more—or raise more capital—to buy a property that is going well, has network affiliation, and good cash flow than to take a chance on a start-up. Again, as in radio, if such a station is in a growing market, chances are that its revenues will grow and its value will increase. The cost of operations will not necessarily be any greater, and presumably the talent that makes an existing station popular will continue to work for a new owner.

The best example that I know of in recent years of an entrepreneur putting together a group to buy existing television stations is Television Station Partners, a limited partnership formed by I. Martin Pompadur in 1983 to acquire, own, and operate (and ultimately, by agreement, to sell) four network-affiliated television stations that had been owned by Ziff-Davis Publishing Company. The stations were in Augusta, Georgia; Rochester, New York; Saginaw/

Flint/Bay City, Michigan; and Steubenville, Ohio–Wheeling, West Virginia; and Marty Pompadur was able to purchase all of them for an aggregate purchase price of $56.2 million, which represented a multiple of eight and one-half times broadcast operating income of the stations for the twelve months ending July 31, 1982, the last full year of operations before the contract of purchase was signed.

Very little of Mr. Pompadur's own money was used to purchase the stations, but if they are successful under his management, he stands to make a great deal. Of course, he had an advantage because from October 1977 to July 1982, he had been the president of Ziff Corporation, the parent company of Ziff-Davis Publishing Company, which owned the stations that the limited partnership purchased. He had also been, from March 1979 to July 1982, the chairman of the board of Ziff-Davis Broadcasting Company, which was responsible for operating the stations. Further, he brought into the partnership Ralph Becker, who had been president of Ziff-Davis Broadcasting and who also had five years of experience as executive vice-president and chief operating officer of Rust Craft Broadcasting Company, which had previously owned the stations. Pompadur had also been at the American Broadcasting Companies for seventeen years in various capacities, including television management, and had served on the ABC board of directors.

Thus Pompadur and Becker knew a lot about the television business and about the four stations they were buying. Still, they needed to raise $64 million ($56.2 million plus working capital, etc.) to make the purchase, and that wasn't such a simple task. They did this in masterful fashion by having a group of banks agree to lend them $38 million (which the stations' physical assets, good will, and cash flow were clearly worth) on condition that they raise the other $26 million from limited partners so that the cash flow produced by the stations could be used to pay interest on the loans and ultimately to amortize the principal. An asset purchase agreement was entered into by the partnership with Ziff-Davis. This included a condition that if the money was not raised within a certain period of time, the deal would be terminated.

The package of printed materials prepared for the offering contained a "private placement memorandum" describing the offering of $26 million of limited partnership interests. It was quite a document, ninety-six pages, with additional appendixes and exhibits. It contained a description of the stations and what they had been doing, financial projections for what they might be doing for the next seven years, and a full rundown on the television business in general and on certain legal and regulatory aspects of the industry. Additionally, it contained a tax opinion from an eminent New York law firm; an expert opinion of Blackburn & Company, Inc., one of the leading broadcasting brokerage firms, concerning estimated fair market value of the stations; and an opinion of Frazier, Gross & Kadlec, a well-respected firm in Washington, D.C., that carries out inspections of stations and gives asset valuations to broadcasting properties. Pompadur was successful in carrying out the offering, borrowed the rest of the funds, had a closing, and now the partnership owns the stations. According to Marty Pompadur in a recent conversation, they are doing quite well.

What Pompadur and his partnership did in purchasing the four stations from Ziff-Davis is essentially what anyone can do in purchasing one or more television stations. The big difference is that you may not know the stations and have the same kind of personnel available as Marty Pompadur did. So it is even more important that you examine in depth the stations you may be interested in and the markets in which they are located. The rest of the process is very much like the purchase of a radio station, only more intricate and with more dollars, more equipment, more real estate, studios, office space, and other details.

Paul Kagan Associates, Inc., whom we mentioned earlier in this chapter, has a *Primer on TV Station Investment* that you might want to look at, which can give you a better idea of what the television industry is like and what the fair market value of stations might be. They also give examples of comparable sales. A lawyer, engineer, accountant, plus one or more persons who have had experience operating a station are essential to making your business plan, negotiating the transaction, participating in the financing and the closing, and in operating the station or stations of your choice.

It's Not Too Late to Buy a Cable System

I don't have to recite for you again all of the figures concerning the presence of cable television to convince you that it is going to be one of the biggest industries around, especially if talented programming people get properly involved and recognize it for what it is worth.

The trouble these days is that it is very hard to find a cable property to buy. There aren't too many franchises left in the United States where building can be done on a scale economical enough to allow the market to set equitable rates and permit the cash flow to come in. The large companies, some of which are having their own financial troubles, either have been buying up systems that are already built and that are showing pretty good results, or else they are selling off some of their systems, large and small, to other large multiple-system operators. Despite the dearth of available cable properties, a few can be found, and I believe that there will be more as the days go by.

Good brokers in the field may be able to direct you to available systems, but the probability is that the bigger and better brokers will be arranging for the big companies to make those purchases. However, you may know of a town where no cable franchise has been awarded or where the successful applicant has not been able to go ahead with building the cable system. If you do, you certainly should evaluate the enterprise, and, if you think that it is worthwhile, try to obtain the franchise or to buy the system.

However, if you aren't already experienced, the first thing you want to do is find someone who has had a lot of experience in the cable television industry, preferably someone who has built a system or at least operated one, and have him or her join with you in your new enterprise, whatever the cost, even if you have to give up a good part of the equity to get him or her.

As I've said, if you are interested in getting into ownership in a particular area of telecommunications and you don't know much about it, study up on it as much as you can. Unfortunately, in the cable area, there really isn't much of a practical nature available to read. You might want to get my first book, *All About Cable,* from a library. (It sells for $70 complete, including the latest update, and can be ordered by mail from the Law Journal Seminars-Press, 111 Eighth Avenue, New York, NY 10011; semiannual updates cost $35 each.) With it, you can study the history of the industry, as well as clue yourself in on the legal, regulatory, and business aspects of the field. Again, Paul Kagan runs seminars on cable and cable investment. He also has a registered newsletter called *Cable TV Investor,* which contains very helpful data on the cable television industry. You certainly will want to subscribe to Titsch Communications' *Cablevision* magazine, now a weekly. More information on these can be found in Appendix A.

Going for the Franchise

Cable television franchising in the United States (and now in some other countries like England) has become a long, cumbersome, costly, political, and often unsatisfying process, and I don't recommend getting involved in it to anyone these days, regardless of size and resources. It is a subject that can be discussed for days without resolution. We are getting into a new phase as cable franchises and contracts come up for renewal. Service may not be as good as promised or hoped for; rates need adjustment; some systems cannot carry all of the available television stations, cable programming, public access channels, and supplemental services like fire, police and burglar alarms, and medical alert systems. They are being asked to rebuild their plants and equipment to offer the public more services demanded by authorities.

Sometimes, however, the situation may not be as bad as you expect. The fact that there is no one else in town to fight you for a franchise may allow you to make a reasonable offer of service and equipment—and make pretty good money if you are able to sell the system to enough subscribers and keep them happy and paying your bills. You also may be able one day to sell advertising on your system, and thus gain revenues that warrant all the trouble you may have to go through to obtain the franchise and build the system.

If you really want to go forward in that area, there are some good lawyers like Fleischman & Walsh (where my friend, Stuart Feldstein, formerly general counsel at the National Cable Television Association and experienced in the copyright fees proceedings, is a partner) or Sol Schildhaus, who used to head the Cable Bureau at the FCC. Both are in Washington, D.C. There are cable consultants like Malarkey, Taylor & Associates, also in Washington, D.C., who can help you out. Unfortunately, too often people who want such franchises choose to go to lawyers in the town or city where they think they have the best political connections. While I can't advise you to forget the politics, don't make it paramount. Instead, make the best possible presentation you can, employ experienced people to help you out, and you will do as well at far less cost than if the process were to go on for many months or years with a resultant waste of much time, energy, and money.

You will find some of the people that you might want to consult listed in Appendix D at the back of the book.

How Others Are Buying Cable Systems

If you want to get into cable ownership by way of buying existing systems that have already obtained their franchises, built at least part of their franchised systems, started their marketing, and obtained some payment of receivables, it might do well for you to emulate a really great guy named Paul Field. Paul was once a television producer at some advertising agencies. He became involved with a very fine man, Jerry Greene, in the media brokerage field. They started early on to specialize in the sale of cable television systems and were going at it quite successfully when Jerry was killed in an untimely accident. Paul tried to carry on the brokerage business but then decided it might be worth his while to try to get into ownership. Again he chose a good partner, a younger man named David Pardonner, who had, as senior vice-president of Cable Operations for Teleprompter Corporation, been responsible for the daily operations of 110 cable television systems serving more than a million subscribers in thirty-three states. Pardonner had started at Pacific Telephone as an engineer and a manager, and he had once been the general manager of a cable TV system in West Virginia.

In 1979, Paul and Dave formed Essex Communications Corporation to buy cable systems and to operate them, hopefully at a profit. They found available systems but needed capital to purchase them. It was at that time that some people on Wall Street discovered the tax advantages of buying and building cable systems, using high initial building costs and depreciation together with investment tax credits for equipment as tax shelters for high-income investors in limited partnerships. Paul and Dave found their benefactors at Oppenheimer and Company, and with that firm's help they successfully raised millions of dollars to buy and operate cable systems.

In June 1980, Essex 1980-81 Investment Limited Partnership was formed, with Essex as the general partner. It was funded with $1.2 million in equity, representing thirty-two units of limited partnership interests of $37,500 each. The partnership then acquired three existing cable television systems in the Allegheny Mountain region of West Virginia and Maryland for $3.2 million. In December 1983, the assets of the partnership were sold for $5.25 million about 12.2 times the prior year's operating income from the cable systems. The cable systems didn't perform according to the Essex projections, but the partners estimated receiving total distributions of $56,000 by the end of 1984, so the internal rate of return will be in excess of 25 percent per annum. Not too bad for a start-up operation!

In September 1980, Essex formed another limited partnership to buy cable systems in Traverse City, Michigan, and to develop other cable systems in the area. This time they raised $1.28 million in equity by selling thirty-two units of $40,000 each. Because of the big recession in Michigan in 1982-83, these systems haven't done so well yet, but bank debt has been refinanced and Essex is hoping for the best. In March 1981, another partnership was formed, this time with $13.875 million in equity, and Essex purchased more systems in Michigan, Alabama, North Carolina, and Tennessee. Again there have been disparities between forecasted and actual results (less actual than as forecasted), but the management remains confident of the ultimate success of these operations.

In May 1982, Essex formed its fourth limited partnership and raised $8.1 million in $100,000 units, with which it bought thirteen systems in North Carolina, Oklahoma, and Florida.

In April 1984, Essex decided to purchase more cable systems. It organized another limited partnership, Essex 1984-1 Investment Limited Partnership, which sold, through Oppenheimer & Company, Inc., $8.15 million of limited partnership interests with minimum investment of $100,000, and it purchased initially three cable systems in Mobile County, Alabama; Hancock County, Mississippi; and Santa Rosa County, Florida; as well as a franchise for a system in Dauphin Island, Alabama.

The Mobile County, Alabama, system acquisition may give you a better idea of how these things operate. The system had been built and owned by Bayou Cablevision Corporation, which also had a franchise for Bayou La Batre, Alabama. The total Bayou System was begun in 1979 and now consists of 150 miles of cable plant, which passes 6800 homes, serving 3850 primary subscribers and 3800 pay cable subscribers. The service area that it covers is primarily residential, and it is inhabited mostly by families living and working around Mobile. There is a new high school complex, and a number of new commercial buildings are going up in the area.

The cable system has twenty-eight channels of programming on it, including various television stations, superstations, health and weather channels, and a religious channel, as well as five pay television services, including HBO and Showtime.

Subscribers pay $11.45 a month for basic programming and $10.95 a month for each pay channel, with discounts if two or more pay channels are taken. In April 1984, 57 percent of the homes passed by the Bayou System were subscribing to the basic cable service, and almost 100 percent of these were taking at least one pay service. All of this was producing $400,000 per annum in operating income at the end of 1983. Essex has estimated that operating income in 1984 will reach $500,000.

Through all of these partnerships and the investments they have made in a number of cable systems in various states, Paul Field and Dave Pardonner have maintained their positions as chief operating officers and as general partners, so that you can probably say that Paul has become a person of substance. There is no nicer guy (and he's also lucky to have a charming wife who has supported him in all of these efforts, including sitting through some rather boring cable conventions). He is deserving of all the success that he may achieve in the cable ownership business. Paul is not just a person to be admired for his courage and tenacity, but an example you might want to follow if you hope to make millions in this area of telecommunications.

A number of other people are doing what Paul Field did in cable, but the task is getting harder as the availability of systems decreases and as the cost of construction and operation increases. Hundreds of millions of dollars are being sought from limited partners across the United States, and there is no certainty that everyone will be able to finance an acquisition even if they have the properties and the personnel available. There is also no certainty that a satisfactory number of people will become or remain cable and pay cable subscribers to make an investment and all the borrowing profitable. Yet I think that there is still a lot of room for the smart people to get involved in cable ownership, and I am certain there is a lot more money to be made in the industry by industrious and creative people who enter the field.

The Mechanics of Purchasing a Cable System

The actual mechanics of purchasing an existing cable system are very similar to what is involved in the acquisition of a radio or television station. However, what you will want to look for in agreeing to a purchase price will be quite different. You should carefully examine the franchises and their demographics, especially the population growth figures and what commercial businesses are in or may be expected to come into the franchise market. You can certainly obtain data on households, household income distribution, and population distribution by age, and you will want to look into what the television viewing habits have been in the area.

You must carefully review the franchise agreement that the seller has, especially the requirements for construction of the cable system and the rates for basic and pay services. Any special conditions put in the franchise agreement have to be looked into, as you may suddenly find that you have to build in some remote area at a tremendous cost just to supply the needs of a minimum number of potential subscribers, and this could cause you to quickly file for bankruptcy.

Naturally, you will want to know how much of the cable plant has been built, how many homes have already been passed, how many remain in the construction phase, how much can be built above the ground, and how much and where must cable be put underground. You will want to know the costs of in-place construction and what kind of revenues have been produced.

Services provided and to be provided and the rates that you will be allowed to charge are probably the most important facts to be determined, but the method of marketing the system and the present and potential competition are also important elements to be investigated.

Assuming that all looks right, you will then have to make a deal with the present franchisee, making certain that you and your company or partnerships can be approved as the assignee by the franchisor. Obviously, no franchising authority that has taken great pains to award the franchise in the first place will want to let the present franchisee assign the franchise to just anybody, so it's very important that this be cleared well before you actually sign anything.

The purchase price generally reflects the number of actual basic subscribers that the existing system has. An industry standard of some dollar multiple of this has been used in the past few years. There have been some cases of people paying up to $1000 a subscriber for the opportunity of taking over a good system with no or little construction problems remaining, good equipment, a long-term franchise, and very good cash flow. I remember when cable was just getting started, so I still think that no one should pay more than $500 a subscriber for a cable system. But there is so much competition going on for any system, no less a good one, that this is probably on the low side these days. Anyone who can buy a system and spend in total no more than $750 a subscriber for the purchase and the cost of rebuilding is probably doing quite well.

Once you have settled on terms, it remains for the lawyers to work out the details, which usually involve an asset purchase agreement similar to that used in broadcasting transactions. The acquisition may take place a long time after signing such an agreement because of the financing condition that is usually put into the agreement, especially if you are using the generally accepted form of limited partnership financing these days. Again, it is important that all the arrangements for taking over control are made well before the closing, especially for billing to subscribers, the lifeblood of the cable system.

Going Further into Ownership—Where's the Money?

It's one thing to have the expertise and the desire to become actively involved in these aspects of the telecommunications industry. It's another thing to have the money to do it. Not all of us have the capital resources to make the plunge. So that means that we link up with associates who have the money to help us along, get some people to invest with us, or get financing from a bank or other lending institution. You probably will have to do *all* of these to do it right. You also have to be creative when it comes to corporate and tax matters in order to obtain the capital to put you into the ownership arena. Let's take a look at how that can be done in the next chapter.

How to Get the Money
to Set Up Your Own
Telecommunications Company

If you seriously want to get involved in telecommunications as an owner, and have the qualifications to do so, don't be deterred by the fact that you don't have what you think is the necessary capital in your bank account. Whether you are a man or a woman, you *can* set up your own communications company (or other appropriate entity) to own and operate radio or television stations, cable systems, telephone equipment stores, paging device marketing operations, programming production studios, or whatever other communications business you are interested in. Within the constrictions necessitated by taking in or borrowing money from others, you *can* be your own boss.

Very often people come into my office with business plans for telecommunications enterprises that they obviously are qualified to start or expand, yet they fear that they can't push because they simply don't have the money. Such a meeting may take place only an hour after a friend has telephoned, asking me where he or she can invest in one of the telecommunications enterprises I'm always talking about. Or the call comes shortly after the head of a communications lending group of a major bank has called me to ask if I know of anyone who wants to borrow millions of dollars to acquire radio or television stations or to make other good telecommunications acquisitions. It is not my business to be a money-finder for anyone. But there's no reason why people such as I can't bring together for their mutual benefit the people who need it and those who have it, as many professionals do every day. A number of investment "intermediaries" are listed at the end of this chapter.

To Get the Money, You Have to Have a Plan

There are plenty of funds out there looking for good telecommunications investments. Peculiarly enough, the major problem is that presently there are not enough really sound deals for investors and/or lenders to become involved with. Many times the good deals can't be found because potential entrepreneurs fail to make the appropriate presentation to those who have the money or who know where it can be found. As I've said before, too many people have wonderful ideas but are unable to put them down on paper. Or they fail to solicit the advice of professionals who can assist them with preparing a *business plan* that includes projected financial statements that allow their story to be told to the potential investors or lenders. This is the *essence* of financing any enterprise or particular transaction.

SETTING UP THE ENTITY—CORPORATIONS AND LIMITED PARTNERSHIPS

You may already have a corporation that has been buying broadcasting properties but want to do another acquisition. You may simply be starting from scratch, in which case I always recommend that a corporation be set up to carry your business plan forward. Then you may want to consider setting up separate subsidiary corporations for each operation that you want to carry on or for each acquisition that is to be part of your total program.

In certain situations, particularly where there may be significant tax advantages offered to investors (and you) through investment tax credits and depreciation of the cost of something that will be built, you may want to consider using a limited partnership. In any event, you will want to determine what the best entity is and how it should be structured. It's simple enough these days to file the necessary papers to form a corporation or to establish a limited partnership, but not so simple using those entities to raise funds. You will want to capitalize and structure them in the best way possible for immediate and future effectiveness. You should consider legal responsibilities as well as tax ramifications.

WATCH OUT FOR THE SECURITIES LAWS

Obviously, if you are raising money from other people, you have to comply with the laws of all of the applicable jurisdictions. In the United States, unfortunately, there are not only securities laws promulgated by the Congress and constantly changing regulations of the Securities and Exchange Commission, but in each state you will find there are securities laws, and these may vary a great deal from place to place. Naturally, if you are looking to raise money outside the United States, you will have to respect each country's applicable laws and the regulations of their appropriate administrative agencies.

Sometimes people become overwhelmed by all the red tape and become fearful about paying fees to communications, corporate, tax, and international lawyers; accountants; consultants; and other people who may be involved with structuring the organization and any deal it will make. But these are things that one must contend with in order to obtain other people's money (especially if it's a lot of money) and to avoid problems that might otherwise lead to legal complications for the organizer.

Where's the Money?

Assuming that you have selected what it is that you want to do and with whom you want to do it, and that you have written up a proper business plan, then what you want to know is, Where's the money? Well, the money is all over the place. It is worth your while to consider speaking with venture capital companies, perhaps in your town or in a financial center, with banks, with investment bankers, insurance companies, government agencies, your lawyer, your accountant, industry associations, even your Uncle Harry or your cousin Theresa, as to where you can obtain the funds to fulfill your plan. You may also want to contact a local stock brokerage firm to find out if they have an investment banking or venture capital department. Ask them if they can set up an appointment for you to talk with someone about your new enterprise.

PROGRAMMING FUNDS

If you want to go into programming, then you may look into the same sources, as well as the organizations where or through whom the programs that you want to make may be broadcast—the networks (CBS, NBC, ABC, the Public Broadcasting Service, and now some of the remaining cable satellite networks, such as the Disney Channel, Lifetime, Home Box Office, Showtime, and even the Playboy Channel); the film companies (Warner Brothers, Paramount, Columbia Pictures, 20th Century-Fox, Universal, Orion, and possibly Tri-Star); larger TV production companies (Lorimar, Goodson-Todman, Reeves, Paramount Television); even domestic and foreign advertising agencies (such as Young & Rubicam and Dentsu, a Japanese ad agency and the largest in the world, which has an office in New York to help produce programming for Japan and elsewhere); and others who may want to share in your potential profits by giving or lending you the funds to make your programs. Producers of programs have made all sorts of deals all over the world to get the money to make a variety of programs; there are many ways in which deals can be constructed so that the producer need use little or no money of his or her own.

A group I know that wanted to make a special television series obtained a commitment from a network to put on a program if it was made. They then took the commitment to a number of persons and organizations to raise production capital. The series was done with funds derived partially from the network and partially from these resources, which wanted to share in the supplementary and residual values of the programs. The backers included Dentsu, which put up several million dollars in exchange for the Far Eastern rights to the programs for television, cable television, and videocassettes only. In that case about $17 million was involved and it took three years to put the package together. (It resulted in the miniseries "Marco Polo," televised on NBC in the United States and then over a number of television channels throughout the world.) I've seen similar situations where only $250,000 was needed to make a program that ran three or four times in the United States and then in many other countries, with new sound dubbed in for language changes.

You may have had only a little experience in producing programs, but a terrific desire to make a particular program. Don't be carried away and try to produce it with your own money or with that of your relatives if you have no idea whether it will ever be put on anyplace. There are ways to obtain the funds, even some from a lender. First you must determine that a person or organization will make a commitment to you to put it on or sell it somewhere.

SEMINARS

Financial seminars on various areas of telecommunications go on all over the world every week. You may very well want to attend at least one of them to learn more about loans and equity money for station owners, cable system owners and operators, satellite master antenna television systems (SMATV), microwave distribution systems (MDS), or direct broadcast system (DBS) operators, programming suppliers, and any other enterprises that you may consider starting or expanding. Educating yourself about some of the factors that go into equity and debt financing before you approach potential investors or lenders will give you a leg up when you go to seek financing. Such adequate preparation is essential. Representatives of venture capital companies and lending institutions attend these seminars, and it is a good place to meet them— and for them to meet you.

BROADCAST FINANCING

If you and your advisers believe that you don't have sufficient equity capital from your own resources or from gathering together investors, then you need to borrow additional money to operate over short or long-term periods or to acquire other already existing telecommunications enterprises. There are now tremendous financing opportunities in the broadcasting area of the telecommunications business. It is again mostly a question of getting to the right people with your concept, either by dealing with a leading bank's communications group or by negotiating with a local bank that may be interested in lending to your enterprise if you can educate their senior officers about the industry and, of course, your particular idea. A number of banks and investment banking firms specialize in financing the acquisition of broadcast properties. At the end of this chapter, I tell you about a few of them and some of the people who work in them.

THE BANKS IN BROADCASTING AND CABLE

It used to be that banks would never lend funds to broadcasters because they were not certain of the longevity of a broadcasting organization when the FCC required radio and television licenses to be renewed every three years. Also, at one point, there was a concerted attempt by some citizens' groups to have license renewal applications denied to some broadcasting licensees who they thought were not acting in the public interest. Nowadays, especially with the high prices that have been paid in recent years by those who want to get into the field, notably by those who want to enlarge their broadcasting operations, many banks and insurance companies are eager to become lenders or to have their loan portfolios grow in this dynamic and very profitable business.

Making the Presentation—the Evaluation of the Deal

In attempting to obtain funds for an acquisition of a telecommunications property, the potential borrower must do the following:

• Present all the facts concerning the station(s), the cable system, or whatever kind of enterprise you are interested in.

• Set out a list of the equipment to be owned, including any transmitters and towers, or head-end equipment, office furniture, microphones, tape recorders, etc.

• Describe the land to be owned or leased, the buildings that may be involved, such as offices and studios, and where the transmitting equipment may be located.

• List the personnel employed and all of the various expenses that may have to be incurred in making an acquisition and in carrying on operations.

• It helps to have a full appraisal done and written up by a qualified company. Give that to the bank along with your plan and projections.

The banker (as does the potential investor) has to look at a number of factors, including:

• Past performance of management;

• The size and location of the market in which the station or stations may be operating;

• The purchase price that may have been agreed upon in relation to the cash flow that the stations have been producing and what may be projected;

• The kind of debt service coverage (which can essentially be measured by taking the cash flow and determining if it will permit the required payment of principal and interest to the bank over the suggested period of years), and how much of the equity is being committed to the enterprise by the principal and his or her associates and any other investors that may be attracted to it.

Generally, no bank will want to work with someone who never has owned or at least been the general manager of the kind of property that he or she is interested in acquiring. Therefore, it is an essential part of any presentation that association with people having good experience in the field be indicated. Most major banks involved with broadcast lending these days are interested only in lending to established group owners. It is incumbent upon the *new* entrepreneur to indicate strongly that he or she *can* (or there is available management who will) carry out the kind of business plan that is presented. It must also be clear that the pro forma projected financial statements are reasonably accurate. As Alan Griffith, senior vice-president of the Bank of New York, who has been very much involved in this field of lending, said recently, "Broadcasting is not a labor-intensive business, but it certainly is management-intensive."

It has become fairly standard that banks will lend somewhere between four and a half and five and a half times a radio or television station's broadcast operating income or cash flow. But the size of the loan being requested is also very important, as is the amount of equity being put in by one or more of the principals. Clearly, if a rather inexperienced group wants to borrow $300,000 to make an acquisition of a radio station and is putting in half of the necessary capital, it is a lot easier to borrow than in a case involving $2 million where only 25 percent is equity and the balance is sought from the bank. Most of these loans are for a term of somewhere between five and ten years, although some insurance companies are known to have loaned broadcasting ventures significant sums to be repaid over a period as long as fifteen years. Again, what kind of cash flow *has been produced* by the entity to be acquired and what kind of cash flow it *is expected to produce* under new ownership are probably the most important factors in determining purchase price as well as the amount to be loaned for an acquisition by any institution.

Financing Cable Systems

Things are a little different in financing cable systems. There, a great deal of financing is in the development stage. The lender will want to review facts concerning the ratio of debt to the equity being put into the venture and the expected cash flow. Very important also are:

- Figures on the market involved;
- The number of miles that have to be built;
- Number of potential basic and pay subscribers;
- Other media in the market;
- Obligations to satellite programmers.

If an existing cable system is involved, you will have to show:

• What the history of the system has been, including all of the mentioned factors;

• Data concerning cost of operations;

• How much more building of the system will have to be done and how much that will cost;

• How much rebuilding of an older system will have to be done and what that will cost;

• The number of existing subscribers for basic and pay service;

• The full potential number of subscribers the system may allow;

• The "churn" rate (how many subscribers ask that their

lease be terminated, both for pay channels and/or basic service);

• How you, as the new owner, expect to operate and make the system even more profitable. Again, your or your associates' experience in managing systems is most important.

Assuming that you prove to the lender that you are worthy and your venture is secure enough to lend to, then how much money you borrow and on what terms may very well require some extensive negotiating. Lenders today are not very eager to make fixed rate loans. Generally they want to tie interest rates to the prime rate. A good long-term loan should not require payment of more than one and one-half percentage points over prime. Any short-term funds that are borrowed should be for no more than one percentage point over prime, because, hopefully, this will be paid back on a revolving credit basis. If you can obtain some kind of cap on the loan, you really should make an effort to do so in these hard times of fluctuating interest rates. This may require a quid pro quo of a basic floor interest rate that you will have to agree to pay even if rates go down significantly, but you should be willing to do this in order to get the ceiling.

Since the highest costs of operating a new entity, even one that is acquired, are in the first two years, you should try to obtain a condition that during that period interest only is paid on the long-term loan and that amortization of the principal should commence only after twenty-seven months of operations. If you can get the lender to agree to some specific payments of principal on a regular basis thereafter and then call for a balloon payment after the long-term period is up, say, at the end of seven years, you (and the lender) will have ample opportunity to review your operations, and you can either renegotiate the loan at a later time or make other arrangements for paying off the loan, if you haven't sold the property by then.

The loan agreement and other documents concerning whatever deal is made with the lending institution are very important, because there may be many restrictions placed on you by the lender that you hadn't considered, which will affect you for a long time to come. These may include a restriction on the amount of any management fee that you may want to take out of the company, a requirement to maintain a certain amount of working capital at all times, and even restrictions on the salaries paid to employees.

You must have your lawyer and accountant carefully review all of these documents with you to decide if the terms are something that you can live with. Similarly, carefully consider the security documents that you may be asked to sign, as you may be mortgaging your whole business to the bank with the proviso that you could lose it all if you miss a single payment! Banks don't want to run anyone's operation, particularly not a broadcasting business. But if you aren't careful with the documents at the outset, you may find that you no longer own your equipment, land, towers, or what-have-you because your business manager failed to make arrangements for a one-quarter payment of interest and/or principal on a long-term loan.

Finding the Equity Investors

In financing transactions, obtaining enough equity capital and arranging sufficient debt financing seem equally important. But you probably will find that if you set out your business plan, determine how much capital you and your associates, if any, have to put in, *then* make your loan arrangements, the balance that you need to carry on can be found much more readily than if you started out the other way.

What you can offer to potential investors depends upon the enterprise and the way you structure the deal. I always think that the entrepreneur deserves at least a 25 percent edge over any passive investors just for setting up the business, finding the people and the properties, or whatever else may be done to get the enterprise going. After that, it's essentially whatever the market will bear.

When we were establishing a corporation to get into the communications business, my colleague and I invested in voting common stock at a certain level of dollars per share, and we then sold some additional voting stock to a third party at a cost (price per share) that was 25 percent above ours. We also sold some nonvoting common stock to the third party at an even higher cost per share than he had paid for the voting stock, and he also purchased a "convertible subordinated debenture." This can be convertible into a portion of the nonvoting stock for the same amount of dollars as were used by him to purchase the debenture. By intricate mathematical calculations, we arrived at a total equity ownership of the parent corporation that gave the two original entrepreneurs (my colleague and me) a slight edge and also gave the third investor a fair percentage of the total equity ownership for the money.

The three of us then had the corporation borrow some more funds from a bank that our third investor was doing business with, and the total capitalization of the original corporation (less some working capital that we left in the parent corporation) was invested in purchasing 100 percent of the common stock of a newly formed corporation set up to be the licensee of a combination of radio stations that we purchased with a long-term loan given to us by a leading bank in the city where the stations are located.

When we wanted to acquire two more stations in another city less than a year later, we set up another subsidiary corporation to be the licensee, invested some money of the parent corporation in the new subsidiary to purchase all of that corporation's common stock, and made arrangements with a local bank in that city to lend the corporation funds to make the purchase. Parenthetically, the bank in the first city was so insulted that we hadn't borrowed the funds for the second purchase from them that we had to ask the second city's bank to let them participate in the new loan, which they did.

To raise the $1 million balance needed to make the acquisition, which was essentially for all cash, we established a convertible preferred stock in the second subsidiary corporation that was sold to seven new investors pursuant to a "private placement memorandum." This preferred stock pays an 8 percent cumulative dividend, and the investors have the right to convert it into what will amount to 25 percent of the outstanding common stock of the parent corporation if they choose. By so converting, these investors will have not only the benefit of gains made by the stations in one market, but could share in the value of the stations in the first city and in any other acquisitions that are made by the parent corporation or any other subsidiary that may be formed.

In our particular situation, we knew the new investors personally (my two colleagues also participated in the new investment at the same price as the seven others), and we didn't have to go to a money-finder, venture capitalist, or investment banking firm to obtain the $1 million. In fact, some of the investors are clients and friends, and I felt secure enough in what we were doing to suggest the investment to them. I also knew that they had enough net worth both to comply with the securities regulations and to afford a loss of their money (albeit not happily) if the investment did not pan out as predicted.

Similar structuring can be done, utilizing various forms of corporate securities to attract investors and still preserve a good position for the entrepreneur, in order to establish a business or to make an acquisition. If you can't construct something yourself, there are plenty of professionals around who can help you out. In certain situations today, government agencies such as the Small Business Administration or economic development agencies in states and local areas can be of assistance. In the field of broadcasting, the National Association of Broadcasting, located at 1771 N Street, N.W., Washington, D.C., has been very helpful in assisting new people, especially minorities, in getting into ownership, and now a funding organization called BROAD-CAP has been established for this purpose.

Lawyers are the best sources of such creativity in organizational structuring, and no matter which way you want to turn, you should consult a good one. Maybe this book sounds like a public relations tool for the American Bar Association, but I believe very strongly in the theory of business entrepreneurs doing nothing without consulting counsel *first*, not later, and hope that you will, too.

Don't Be Ashamed to Copy from Prior Deals

You should never ignore history in doing anything in life; we can always learn something by reviewing what people have done before us, trying not to echo their mistakes. This certainly applies in the matter of financing in the telecommunications business. There have been some very sophisticated transactions constructed in the last few years to raise capital, both publicly and privately, for buying television stations and cable systems; setting up SMATV, MDS, and DBS operations; establishing cellular radio-telephone and paging device businesses; and others.

You shouldn't hesitate to get hold of whatever written materials you can on these deals so that you can either follow their structures to the letter or at least adapt them for your own purposes. The brokerage houses are the best sources for obtaining these materials, reporting both public and private transactions. Law firms and accounting firms usually have materials on previous transactions they have handled or which they may have been asked to examine by their clients. Sometimes it is worthwhile to go to the local Securities and Exchange Commission office, or the main one in Washington, D.C., to examine public records, or even to the state capital, where there may be securities filings for you to look at.

Sometimes you will like what others have done to raise $10 million, and you may want to use the same structure to plan to raise $500,000 for an enterprise you want to start on a smaller scale in the same telecommunications area, especially if what has been done has proved successful. The same principles may apply, and you may like what was done so much that you will want to follow it exactly for your own enterprise. However, every deal is different, and you probably will create your own distinctive structure anyway, even if you build upon something that was done in the past.

For example, in recent years the limited partnership structure has been successfully employed in raising a lot of money for building and buying cable television systems. Marty Pompadur, whom I mentioned in connection with his acquisition of the Ziff-Davis television stations, thought that the best way to raise $26 million to obtain the $38 million of bank financing he arranged was to use the cable limited partnership form in order to flow net operating losses directly to equity holders. His Television Station Partners anticipated such losses for the first three years of the partnership, and he attracted the investors by offering them (all of whom presumably could use such losses to offset profitable operations elsewhere) these losses.

Profits and losses will be allocated 99 percent to the limited partners until such time that the capital accounts of the limited partners, without giving effect to distributions, equal 150 percent of their investment. Thereafter profits and losses will be allocated 65 percent to the limited partners.

Someone else might not want to do what Marty Pompadur's limited partnership had to do to raise such funds, especially since Marty promised to operate the stations at maximum profitability and to sell them off in a certain period. I'm not sure that on hindsight he would use the same structure today that he used then (as against perhaps the use of a corporation and some kind of convertible preferred security or a debenture with a warrant on some common stock) and with the same restrictions and conditions. Presumably he did then what he thought he had to do to establish a television station–owning venture. Today, people have the benefit of looking at the vast amount of paperwork that was so excellently prepared for Marty Pompadur's transaction and adapting it for their own requirements, again using good professional assistance.

There are a number of similar situations in the cable field. If you are interested in obtaining financing for cable, you should locate and read the private placement memorandums obtainable from firms that have been the selling agent for the transactions, or even possibly from investors (for example, the Essex Partnerships mentioned in Chapter 3). For deals that have been fully registered for public sale under the securities laws, the documents should be readily available. Of course, just because one enterprise sought to go the public route does not mean that you have to do so, but you may be able to adapt a publicly owned entity's structure for a purely private fund-raising transaction.

In this age of tremendous free enterprise and hopefully continuing strong economies, you can start a telecommunications venture of your own even if you lack immediate capital. You can do it and you should try to do it, if that's what you truly want to do. You may not find the money to carry it out the day after you start looking for it, but you can and you will sooner or later if you persist in doing so. I'd like to believe that we will enjoy better broadcasting, stand a better chance to see a potentially valuable cable television industry work, develop some effective and needed educational programming, fulfill the world's everyday communications and informational needs, and perhaps make a far better life for many more people in the world *if* you do. Even more practically, and without being crass about it, you might make a lot of money for yourself, your family, your investors, and your lenders and thus make your life and theirs a lot better because of it.

Financing Sources

To make it a little easier for you to locate sources of financing for a telecommunications enterprise you may be interested in starting or having grow, there follows a list of banks, investment bankers, and venture capital companies that may very well satisfy your need. Write or call them when you are ready, and remember at least some of the things that I have said here when you are talking with them about fulfilling your needs. Don't forget that they probably want to lend their money to you and maybe invest it with you as much as you want to get the money from them.

BANKS

Irving Trust Company
Communications/Media Group
Commercial Banking Division
245 Park Avenue
New York, NY 10167
(212) 408-4740
Ms. Pat Tresnan, Vice-President; Ms. Katherine D. Lorenz, Lending Representative; Mr. John Hall, Assistant Vice-President

Chemical Bank
277 Park Avenue
New York, NY 10172
(212) 310-6161
Stewart Cahn, Vice-President of the Entertainment Group

The Bank of New York
51 West 52nd Street
New York, NY 10019
(212) 536-4903
Alan Griffith, Senior Vice-President

Bank of New England
28 State Street
Boston, MA 02109
(617) 742-4000
Ms. Katherine Marien, Vice-President

Union Bank of California
445 South Figueroa Street
Los Angeles, CA 90071
(213) 236-5000
Craig Dougherty, Vice-President Western Region

AmeriTrust
900 Euclid Avenue
Cleveland, OH 44101
(216) 687-5000
Chesley Maddox, Commercial Bank Officer

Fleet National Bank
111 Westminster Street
Providence, RI 02903
(401) 278-6000
John Barber, Vice-President—Commercial Bank Officer

Bankers Trust Company
280 Park Avenue
New York, NY 10017
(212) 775-2500
Will Archer, Vice-President of Media Division

Marine Midland Bank
140 Broadway
New York, NY 10015
(212) 949-6500
Jose Echevarria, Vice-President Broadcasters Department; John
Brooks, Commercial Banking Officer—Broadcasters Department

Republic Bank
P.O. Box 225961
Dallas, TX 75265
(214) 922-5000
Ms. Phyliss Riggins, Vice-President

Citizens Trust Company
146 Westminster Street
Providence, RI 02903
(401) 456-7000
Thomas Hoagland, Senior Vice-President of Corporate Banking;
Robert Sanderson, Senior Vice-President of Corporate Banking

Norstar Bank Upstate
69 State Street
Albany, NY 12201
(518) 447-4109
Edmund Duffy, Assistant Vice-President—Commercial
Division

Canadian Imperial Bank of Commerce
Corporate Finance Group, U.S.A.
30 North La Salle Street
Chicago, IL 60602
(312) 726-8858
Barton T. Schneider, Senior Manager

INVESTMENT BANKERS AND OTHERS

TA Associates
45 Milk Street
Boston, MA 02109
(617) 338-0800
David Croll, Managing Partner; Richard Churchill, Jr., General
Partner; William Collatos, Vice-President

Hartstone & Dickstein
60 Washington Street
Hartford, CT 06106
(203) 549-7000
Barry Dickstein, President

First Chicago Investment Corporation
1 First National Plaza
Chicago, IL 60670
(312) 732-5400
Paul Finnegan, Investment Manager

Communications Equity Associates
851 Lincoln Center
5401 West Kennedy Boulevard
Tampa, FL 33609
(813) 877-8844
Harold Ewen, President; J. Patrick Michaels, Chairman of the
Board; John Long, Vice-President

Daniels & Associates
2930 East Third Avenue
Denver, CO 80206
(303) 321-7550
Phil Hogue, Executive Vice-President of Investment Services;
Jay Busch, Senior Vice-President; Timothy David, Senior Vice-
President

VENTURE CAPITAL FIRMS*

Adler & Company
375 Park Avenue
New York, NY 10152
(212) 759-2800
Frederick R. Adler, Managing General Partners; Joy
London, General Partner

R. W. Allsop & Associates
Corporate Center East, Suite 210
2750 First Avenue, N.E.
Cedar Rapids, IA 52402
(319) 363-8971
Robert W. Allsop, General Partner; Paul D. Rhines, General
Partner

*Courtesy of *Venture* magazine

Burr, Egan, Deleage & Company
3 Embarcadero Center, Suite 2650
San Francisco, CA 94111
(415) 362-4022
Jean Deleage, Shirley Cerrudo, Tom Winter (partners)

Burr, Egan, Deleage & Company
One Post Office Square, Suite 3800
Boston, MA 02109
(617) 482-8020
Craig Burr, Bill Egan, Brian Applegate (partners)

The Centennial Fund
600 Cherry Street, Suite 600
Denver, CO 80206
(303) 329-9474
Steven Halstedt, General Partner; Jack Tankersley, General Partner; Larry Welch, General Partner

Ferranti High Technology
505 Park Avenue
New York, NY 10022
(212) 688-9691
Michael R. Simon, Vice-President of Venture Capital

General Electric Venture Capital Corporation
3135 Easton Turnpike
Fairfield, CT 06431
(203) 373-2154
Harry T. Rein, Chief Executive Officer & President; Stephen L. Green, Vice-President & Treasurer

Kleiner, Perkins, Caulfield & Byers
4 Embarcadero Center, Suite 3520
San Francisco, CA 94111
(415) 421-3110
Eugene L. Kleiner, General Partner; Thomas J. Perkins, General Partner; Frank L. Caulfield, General Partner

Michigan Capital & Service, Inc.
500 First National Building, Suite A
Ann Arbor, MI 48104
(313) 663-0702
Anthony F. Buffa, Vice-President; James A. Parsons, Vice-President

Rothschild Ventures Inc.
One Rockefeller Plaza
New York, NY 10020
(212) 757-6000
Robert S. Pirie, President of Rothschild, Inc.; Archie J. McGill, President of Rothschild Ventures Inc.; James C. Blair, Managing Director of Rothschild Ventures

The First Boston Corporation
Park Avenue Plaza
New York, NY 10055
(212) 909-2000
Denis Newman, Managing Director in Investments; W. Barry McCarthy, Jr., Assistant Manager in Tax Exempt Sales; Harold W. Bogle, Assistant in Corporate Finance

Creating Programs
to Make Millions

It is abundantly clear that you can't use any of the amazing telecommunications technology without programming. This is true even to some lesser extent for use of telephones, telex, and similar equipment, because the user makes his own "software" when utilizing such technology to convey information from one place to another. It certainly is true of other technology, such as radio, television, cable and pay television, multipoint distribution service, satellite master antenna television, satellites, and computers, which require programs for people to hear and/or see, or to at least operate the system. You can have all the hardware in the world and use the most current technology devised by engineers, but if you don't have the software that a lot of people want to see or must use, your hardware is going to be worthless. The owners of such facilities never would be able to operate them if it were not for the creative talents of those who write, compose music for, act in, direct, photograph, record, and produce the programs that are heard and seen by literally billions of people throughout the world every day.

Lots of people are making millions in programming. Television and radio networks and stations are paying millions of dollars a week for programs, in order to sell the advertising that accompanies them where potential customers can see it. As cable systems try to recruit and keep subscribers, and as videocassette stores proliferate in an attempt to sell or rent more prerecorded programs to a populace that has gone wild for VCR equipment, the demand for programs increases, and the money to be used and made in producing and selling such programs multiplies. Some producers are making $100,000 *a week* for putting programs on national networks. Johnny Carson actually gets, in one form or another, $8 million a year for being on "The Tonight Show" on NBC four times a week for approximately forty weeks a year. A good actor on a long-running television series can earn $50,000 a week for doing whatever he does so long as his popularity continues and the show is on the air.

As the use of radio expands, the need for radio programs that attract and hold listeners also expands, so that advertising time can be sold nationally, regionally, and locally. In certain cities, radio personalities are paid more than $1 million a year for attracting and keeping an audience of devoted listeners. Good programming consultants, the people who put together the music that is played on stations throughout the country, can earn over $100,000 a year. Paul Harvey, whose daily news and commentary programs are carried on over 200 radio stations across the United States every weekday, is probably the highest earning individual radio broadcaster in America today. He is reported to earn well over $1 million a year.

All that programming requires many people—disk jockeys, moderators, actors, boom men, editors, line producers, news readers, news writers, etc.—and certainly the engineers who make sure that it all goes on. There are literally hundreds of jobs in the electronic communications field in one small city alone. You can imagine what the need for people is in places like New York and Los Angeles, and other major world centers.

You Have to Believe in Your Talent

While everyone looking in from the outside thinks it is impossible to get in on any of this, at least in any *big* way, the fact is that there is a constant search going on by owners and operators of telecommunications properties for talent and programs. The rewards for getting on yourself or having a program that you developed put on television, radio, cable, videocassettes, or some other telecommunications technology are rising rapidly. So is the competition. But if you really believe that you have the talent, go for it.

The most important thing is to make a decision about what you think you can do. Maybe this has come through college training in a communications program, or even in some engineering courses. It may involve obtaining a job at the lowest possible level in a local radio or television station or with one of those new cable systems that have gone up in your area. Perhaps you have a talent for writing, which you haven't used for years; the children have grown up, and you are looking to make your contribution to the Information Age. Perhaps you need to collaborate with someone who has had some experience in writing for television, so that you can create a new "Dallas" or "Dynasty." It may be that you are a reporter with a local or area newspaper and you're dying to get into television news. You believe that you are attractive enough to be an anchor person on the six o'clock news that you keep watching every night but can't stand. Maybe you have a great idea for a cable television series or for some music videos that you are sure will go over well on HBO or MTV, making them (and you) at least a million dollars if you can only find the money to have it produced.

YOU HAVE TO THINK POSITIVELY

I'm not Dr. Norman Vincent Peale and I don't have any secret for positive thinking, but if you truly think that you can do something in some aspect of the telecommunications business, you should have confidence in yourself and make the effort to bring it about. If it requires going to New York City, Los Angeles, London, Tokyo, Caracas, Copenhagen, or wherever to get ahead in what it is that you want to do, then go! Sure, the odds are very high against you. But you may make it, and if you do, you have a chance of vast success in the field, not only from the point of view of satisfying your ego, but in terms of tangible wealth as well. There are and will be more dollars earned for successful programs in telecommunications than ever before, and there is no reason why a lot more people can't share in it.

I recently had separate conversations with two good friends who have done pretty well in the television programming business—Giraud Chester, who is executive vice-president of Goodson-Todman Productions, and Edward Bleier, who is executive vice-president of Warner Brothers Television, a part of Warner Communications, Inc. Goodson-Todman produces some of the leading game shows on U.S. television, including "Family Feud"; they presently have ten game shows on television in England. Warner produces and distributes a number of weekly dramatic and comedy television programs and some specials. Both Jerry and Ed have been around the programming business for a good while, and I asked them what they thought of the future prospects for people who want to get into the programming business.

Both Jerry and Ed told me that the prospects for producing programming and making a lot of money at it are greater than ever, but that the competition is even more fierce. They say that the most important thing, if financial success is your goal, is to have a program produced and put on the air. There are a number of people who will tell you that often this involves not simply utilizing talent and original ideas, but whom you know, and this may very well be true.

While cable television looks like a rapidly rising industry, the truth is that comparatively little money is being spent by the cable people in producing original programming. However, both Jerry and Ed said—and I agree—that if a program can be put together that has a number of markets, you may very well create possibilities for distribution in several broadcast areas, including television, cable, and videocassettes in the United States and throughout the world. You could make a fortune if all that came about. This also works the other way. For example, Radio Caracas International, in Venezuela, produces programs put on in that country and then sells them to be televised in the United States and other Spanish-speaking markets.

YOU HAVE TO BE AGGRESSIVE—AND BE ABLE TO ACCEPT REJECTION

You have to be aggressive about getting to whoever makes the decisions, regardless of whether you are trying to sell a program in New York or Kansas City, in Caracas or Tokyo, and you have to be able to accept rejection from whatever source it comes. If you really believe in yourself, and the work that you are doing, then you have to keep coming back, trying again and again until you succeed. Almost everyone who has made it has been through this mill, so you don't have to be embarrassed about your aggressiveness and persistence. If one script is turned down by producers or networks, try writing another. If you are rejected by a personnel director or a program manager when seeking a particular position, be willing to accept another position, and don't hesitate to make your willingness known. Sooner or later, if you have the talent, it is going to be appreciated, and when it is, you really are going to enjoy it and the money that you earn!

There is no way in which anyone can encompass in one book all of the places and all of the people who can help start or further your career in telecommunications. If you are trying to produce programs, then you must bring your idea or script to the attention of a network programming department or a production company able to get your program on the air, whether through some kind of syndication deal, on an independent network of television stations, or via a pay cable television system such as HBO, Showtime, or the Disney Channel. Try to find out who the person or persons are that pass upon these matters, then do your damnedest to get your material read. Always ask who makes the decisions, and somehow present your creation to him, her, or them. In New York or Los Angeles, this probably means using an agent or a lawyer to get the material read. If that someone likes what you've written, your agent or lawyer, or both, will make a deal. If not, that's just one more rejection. Keep at it.

Minority Involvement

Minorities in the United States have rightfully been complaining about the failure to involve them in the writing and production of programming geared toward women, blacks, Hispanics, and other minorities—and the general public as well. Unfortunately, the probabilities are that those who have control of the purse strings will continue to try to put on programs that will appeal to the greatest number of people at a time, so members of minority groups who want to succeed in programming will have to work hard at breaking down the barriers.

There is one notable exception with respect to opportunities for minority involvement. It is a great time for Hispanics in broadcasting. Hispanics can put together terrifically salable programs to run on the Spanish Information Network on cable and the independent Spanish-language television stations that are doing so well in the United States. Such programs can also be sold in every Spanish-speaking country in the world, particularly places such as Venezuela, the Dominican Republic, and Spain. There are now about twenty-five Spanish radio stations around the United States (and many more throughout Central and South America and the Caribbean), and they all need people and programs.

MORE MINORITY CONTROL

Those who control the process make the programming decisions. Thus there must be a greater influx of minority ownership, of advertising directed to minorities, and of minority people in executive positions in the communications industry to assure that more programs are written for, produced by, and feature members of minority groups. Too many people have been stuck on the idea that programs must be put on for the masses only, and not very much has happened to alleviate this situation. Change will occur only when there is an appreciation for quality programming that can be disseminated to whatever audience will be appreciative of it, and where the numbers of viewers will still allow the producer to make a profit. Once it is fully built, cable, with its numerous channels and at least one-way addressability, will fulfill this need, but that is still in the future. I still think (as I have prognosticated before) that the due date for really open cable programming is 1990. Meanwhile, people who can create and produce programs must use every method they can to sell their wares now.

Women in the Media

The need for women's programming on television and radio, as well as on the cable television networks, is a rapidly growing one, and the need for women in broadcasting is growing, too. While there have been cases of sexual prejudice and harassment, the obvious need for women in the media and the growing involvement of women in the telecommunications business indicates to me that women are going to be able to find many more jobs in the industry.

In order to show what you can do, you must be creative and innovative in bringing your skills to the attention of the manager or the program or news director of the station or network that you want to employ you. Or you could start in an "interim" position in one of these stations, but keep your eye on what it is that you ultimately want to do. The competition is awesome, but the growing need for talented people should work to your benefit if you get in and do a good job.

Copyright Everything

One word of caution from a practicing lawyer. Always put a copyright notice on any piece of work that you submit to anyone else, even if it is just a concept for a television series or program. Generally, ideas alone are not copyrightable, but if your treatment, outline, or script is comprehensive enough, it may be registered. At the very least, you may have what is known as a common law copyright in what you have written, and it will afford some protection for you. Someone seeking to pirate your treatment will think twice about it if you put a copyright notice on your cover; that notice could serve as some basis for an infringement action if it comes to that. In the United States, you don't have to register your program or idea with the Copyright Office in Washington until it is actually produced and "published," that is, put on the air.

To give yourself some common law copyright protection, you should put on the cover of whatever it is that you have put down on paper a little c in a circle (©) and print next to it "Copyright 198_" (whatever the date of the publication is) and then your name or the name of any company that you may be operating under. Give a copy of whatever it is that you have to a friend or your lawyer or anyone else you trust, and you *may* be deemed to have "published" it if there is any later dispute about someone stealing your idea or work. This should give you some advantage if anyone attempts to appropriate your idea or whatever it is that you have written and you subsequently want to sue.

Funding to Produce Programs

If you are intent on producing a program or even a series of programs, you not only have to have the place to put them on, but the funds to make the program. There are all kinds of sources for this, and we have addressed that problem in Chapter 4.

Educational Programs

You may want to look into, or you may already be involved with, an area of programming that I think has the greatest potential in the United States and elsewhere—educational programs. For some time, I have maintained that if we interpret that word *educational* in the broadest sense to include information and marry it to telecommunications technology, we will have one of the most lucrative parts of the entire telecommunications industry in the not-too-distant future. This will be especially true when cable television is more prevalent in this and other countries. Then people will be able to select their information and their desired course material individually and pay for what they want to study monthly on a selective basis. This will require a much greater expansion of the use of computers and what are called *addressable converters*, so that the cable subscriber can select what he wants when he wants it and pay only for that particular program, course, lesson, or piece of informational material.

Such educational programming using telecommunications technology and computers requires very special preparation and a good deal of research and development. Those who know how to do this will someday, in my judgment, make a great deal of money. It is not as easy as it looks, and for many years a lot of people have poured a great deal of money down the drain trying to perfect informational television in conjunction with all sorts of equipment. Some companies have had success in the field (see Chapter 2), but it will be a few years before any really widespread profits are derived from it. Right now the cable television companies, which should be spending the money and hiring the people to establish educational programming, are caught up in building and marketing their systems; most are reluctant to commit themselves to this form of programming at this time. Hopefully, before the decade is over, they will realize what a tremendous business it will be. In the meantime, those who have the ideas and the ability to bring them to fruition should be gearing up. If you have the abilities and interest, you should be trying very hard to research and develop this kind of programming; you are likely to make a lot of money eventually if you do.

Commercials Can Be Lucrative, Too

If there is going to be growth in the telecommunications programming business, there is surely going to be growth in the amount of advertising on television and radio, including cable television. Advertising requires people to make the commercials.

Directing and producing commercials is an art form in and of itself. We all have noted marvelous examples of great photography, set design, singable music, and even wonderful acting in commercials. Especially impressive is the extraordinary ability of those responsible for them to get across a clear message to millions of people in the course of just thirty or sixty seconds.

From direct experience in representing commercial makers and their corporations, I can tell you that there is a great deal of money to be made in commercials these days—by the directors, producers, cameramen, actors and actresses, announcers, and the many others that it takes to make just one commercial. An advertiser may pay $100,000 to a production company to produce a commercial in three days. Some outstanding directors make $7000 *a day* directing commercials and may go all over the world (and sometimes with very attractive people) to do them.

I have a cousin, Harry Hamburg, who is one of the leading commercial directors in the business. He started as an artist and then went into still photography in New York. He did quite well shooting advertising campaigns for the leading agencies. Harry has a marvelous sense of humor. He used some really funny pictures to sell Volkswagens and other products in the 1960s. When it looked like making television commercials was going to be a lucrative business (and probably more creative than still photography), I persuaded Harry to make the leap. It wasn't easy, especially since he was doing so well in the still business. I made a compromise agreement with him. I set up a separate corporation through which he could direct and produce commercials while keeping up his work in the still photography business.

Harry had a very good sales representative, an ambitious young fellow only twenty-four years old, who was doing extremely well taking Harry's photography portfolio around to advertising agencies and booking jobs for Harry to shoot. I convinced Harry to make up a reel of film material and let Jim, the sales rep, take it around with a request that Harry be allowed to bid on television commercials, especially the funny ones.

In the commercials business, the art directors in the ad agencies make drawings of what different segments of a commercial will look like, with whatever dialogue there will be. These are called "story boards," and when they are ready, they are sent out to production companies and directors to solicit a bid on how much it will cost to make the commercial or a series of them. Jim went to all the agencies he knew in the still business and virtually begged the art directors for story boards to bid on. He received a few after a lot of effort and a great deal of promotion of Harry's company. Harry made some good bids and was finally awarded contracts for some Alka-Seltzer and other commercials. He did an excellent job, and from there on he had it made.

Harry thrived and finally gave up doing stills when it became too much of a burden to do both. There was a period in his life when Harry abandoned all worldly things (he had bought an apartment in the Dakota, a large, old building in Manhattan, which he unfortunately sold for about $800,000 less than Yoko Ono paid for it a few years later) and moved his wife (who fortunately had very wisely invested his money for him) and three children to a hillside in northern California. However, he now has his nose back to the grindstone, working out of Los Angeles and directing commercials all over the United States and Canada. He does very well and seems to enjoy his life a lot. Recently, in one of our transcontinental telephone conversations, I asked him if he would recommend that others go into the commercial production business. Harry said yes quite emphatically, although he also said that at his age he doesn't welcome the competition.

ACTING IN COMMERCIALS

There is another very rewarding aspect to the commercial advertising side of telecommunications. The television networks must pay those who appear in national commercials what are called "residuals." These are fees paid the actors and/or actresses when a commercial airs, and of course commercials air repeatedly. An actress who may speak only one line about her favorite laundry soap or toilet bowl cleaner receives a basic fee for appearing in the original production, then also earns thousands of dollars a year from the residuals if the commercial appears often enough. There are people making $500,000 a year these days acting in commercials, and that's not counting celebrities, who are making even more.

Naturally, competition in the commercials business is really tremendous, but people do keep coming in and up, and the advertising agencies, production companies, and casting directors are always looking for new faces. Nowadays, commercials are made locally and regionally, not just in the big cities, which gives a lot of people more work and the experience to qualify for bigger and more lucrative jobs in major cities here and abroad.

Music Videos

Through the commercial business has come the new phenomenon, music videos. Originally, these were done on the cheap side by rock record companies utilizing commercial directors and their facilities simply to promote records. Because most of these promotions were only two or three minutes long, the commercial directors, who were used to doing thirty- and sixty-second spots, were perfect for the music video jobs. When Warner Communications decided that it might be a good idea to put a music channel on cable television, they were able to persuade some of their record divisions (notably Atlantic and Geffen Records) to give them some of these videos to put on free. They never expected them to be as popular as they have become. Now MTV is an independent publicly owned company, and others are attempting to put on a competing network for music.

Some of the videos now appear on commercial television, and the most successful ones, like Michael Jackson's "Thriller," have become extraordinarily lucrative properties as videocassettes to be purchased in video stores. Directing and producing these music videos is but another example of the need for programming in the growing telecommunications field. Again, it is not simply the idea that counts, but the execution. People who have the talent to make these programs will reap many benefits in the future as they make their abilities known.

Sports Programming

Another constantly growing field for on-air personalities who often end up making a great deal of money is sports programming. Not all sports announcers are ex-players or coaches or managers. The lovable Howard Cosell is a former lawyer. There is a growing emphasis by the networks, local stations, and now cable networks, such as ESPN and USA Network, on sports programs. This creates a great need for people, including women (who need not just announce the more popular women's sports events), to be on the air. It clearly is impossible for all aspirants to be announcers on "Monday Night Football," or to be the replacement for Phyllis George on "The NFL Today," or even to be a multi-talented Vin Scully or Pat Summerall. But there will be a continuing need for men and women on local stations to report the sports news and *local* sports events as they occur. Once you have obtained the experience of handling sports programs locally, you may very well be able to seek the bigger jobs in the larger markets and ultimately end up with one of the national organizations and a large income.

Marv Albert is one of the hardest-working sportscasters in America. He reports the sports news every night at 6 and 11 P.M. on the local NBC station in New York. He also does the play-by-play announcing of the New York Knicks basketball and New York Rangers hockey games on radio and occasionally on television. On weekends, he announces NFL football games on NBC. In his spare time, he handles college games and other sports events on radio and television. Marv reportedly earns over $750,000 a year, and you can see how he might. He grew up in Brooklyn, testing his skills at make-believe announcing of local events such as the stickball and basketball games he played in the school-yard, and then graduated from there to very minor announcing jobs until he caught on. He never refused an assignment. As sports coverage grew, so did the need for Marv's skills, and now he doesn't have enough hours in the week to do all the things that he could.

Don Criqui is a big, handsome fellow, a Notre Dame graduate who works as the sports director at WOR-AM in New York. He announces the sports results about ten times each weekday morning on the "Rambling with Gambling" program, starting at 5 A.M. He then works on weekends as a football announcer for NBC. Don has also handled the tennis matches at Wimbledon and announces a number of college and NFL football games on national television, including some of the Notre Dame games. In addition, you may know Don's face and/or voice from the TWA Airlines television and radio commercials that he does so well. I don't know exactly what Don earns annually from all of these activities, but I imagine that it is a very comfortable sum. He obviously didn't get to that overnight, but started out in the lower echelons and worked wherever and whenever he could. He perfected his mellow voice and his knowledge of sports and kept practicing what is now a masterful delivery. It took a long time and obvious hard work for Don to achieve his success in sportscasting, and he deserves every cent of what he earns. (He also needs it to feed those five great children he and Molly have.)

Clearly, none of the people like Howard, Joan, Marv, Don, or any of the thousands of persons that you see and hear on national or local broadcast outlets just walked in and got the job. I'm sure you will find that hard work and persistence, plus some talent, went into getting each where he or she is today. The future is brighter than ever before in the telecommunications field for people with such qualities, and it doesn't matter what your sex is or how old you are (look at Clara Peller and her "Where's the Beef?" commercials). So stop wasting time, get in on this revolution now!

Getting in Front of the Camera or Behind the Microphone

Ask those who are on radio or television today how they did it, and you'll probably get an answer like "Pure luck" or "I was in the right place at the right time." It probably is true, because that is what has happened to so many people involved in broadcasting, especially those who actually are on the air. However, you (as probably most of those who have succeeded did) really have to work hard at positioning yourself in the right place at the right time. You have to do everything that you can to prepare yourself to seize the opportunity when luck comes about.

I'm very glad to tell how my wife, Joan, got into broadcasting. While at Barnard College, she coauthored a terrific guide book called *New York on $5 a Day*. Joan then tried the theater (unfortunately, I still haven't seen her on the stage), but it seems her persistence in seeking roles wasn't fully rewarded. So she became involved in the advertising and public relations business while still doing more research on the city and revising the New York book every other year, and she wrote shopping and travel articles for *New York* and other magazines in her spare time.

Joan could be as entertaining and informative while speaking as she was in writing, so somebody suggested that she try the radio business. She balked a number of times, but one summer when we were vacationing on Fire Island with our very young children, she put her special knowledge of the city to work and practiced every evening on a tape recorder, doing interviewing and short, informative talks on places and things to do in New York. Joan started to enjoy it and even to think that she might be qualified to be on the air. However, she was busy with her jobs and with raising our two kids and didn't have the time or inclination to go around to stations job hunting in radio—or at least so I thought.

Joan had indicated her interest in going on the air to some professionals, however, and soon was chosen from about two dozen much more experienced or prominent contenders to take over a two-hour-a-day call-in program on WMCA, in New York. Through a change in command after she was there for almost two years, Joan was let go on very short notice. She was traumatized for a while, but she really thought that she was good on the air and that she had something distinctive to offer. She persisted in talking to many people in the business about being allowed to talk about what she really knew—New York. Just as she was ready to give it all up and go back to public relations, Herb Salzman, who was then managing WOR-AM in New York, one of the oldest and most respected (and probably most profitable) radio stations in the United States, called her to say that he was looking for a woman to do some short pieces every morning on "Rambling with Gambling." He had heard her on WMCA, and he wanted her to do some audition tapes.

John A. Gambling is the likeable, very professional second-generation host of the program, which is on WOR every morning except Sunday, from 5 to 10 A.M. It is probably the most successful continuing local radio program on any station in the United States. Since his father, John B. Gambling, started the program more than sixty years ago, there had never been a woman on the program on a regular basis. John liked Joan and what she did, and he agreed to have her on to talk about special aspects of the New York scene, including consumer information, once a morning for about five minutes.

The appearance on the "Rambling with Gambling" program evolved into three pieces each morning, starting as early as 5:20 A.M. The subjects were expanded to cover not only information on shopping, best buys, restaurants, and special daily events around town, but discussions of new products, travel advice, museums, and film and play reviews. Because of her audience acceptance in the morning, Joan was given her own one-hour (now ninety-minute) interview and call-in program each afternoon on WOR. Recently she has also been appearing on the local CBS television station's evening news on a regular basis as a consumer expert. She still manages to find some time to write books and articles and give speeches, and to take care of the children (a little older now) and me. Joan is being paid quite well (though she says that she can always use more money). Most important, she really enjoys her work, although her daily schedule is an exhausting one.

I think that Joan's story should be inspirational for people who want to become involved in broadcasting as performers, especially women, and I would tell it here even if she weren't my wife. She decided on what she wanted to do, packaged her specialized knowledge and talent (she did take some speech lessons from a professional teacher before she went on the air), and went after it. She has been helped by the fact that she has a very believable quality in her voice—something that is very important in broadcasting—and seems to be a *friend* to all her listeners and viewers.

She didn't let adversity stand in her way. I might add that everyone who knows her loves her, including those who just hear her voice, so her persistence hasn't alienated anyone. She has in a way been a pioneer, not only by being the first woman regular on "Rambling with Gambling," but by taking her best asset, her specialized knowledge of New York and its environs, and doing some different things that no one else anyplace in the country had ever done on radio in the same way, to the best of my knowledge.

You, too, can do what Joan did, maybe even better and perhaps more easily. If you want to get in front of the camera or behind the microphone, and you're not just another pretty or handsome face, take your most distinctive abilities, package them to make yourself stand out from the ordinary, and do it! You may know a lot about whatever city or town you live in and can talk about what is going on. You may know about medicine or social services or psychology, or whatever it is that people in your area may not be hearing about, and you know that you can be entertaining as well as informative.

If you aren't in New York or another large city where there are many television and radio stations, then either go to such a place or, better still, try to fulfill what you see as a need in your city or town and get some experience on a local station. You might want to try for a job in a station somewhere else in the country by making audition tapes and sending them out to station managers and program directors or by advertising in one of the trade magazines. You might also speak to one of the placement agencies that advertise in *Broadcasting* and other places. If you are good, someone will find you and bring you on to better and bigger things, if that is your aim. You may be able to get a good agent to represent you and find you that *really* big job. However, there are many people making a pretty good living these days as newscasters, disk jockeys, or interviewers in areas outside of the biggest cities. Some undoubtedly enjoy their lives even more than they would in a highly competitive urban center.

Using an Agent and/or a Lawyer

A lot of people want to know if they have to have an agent in order to work in telecommunications. Since there aren't too many talent agents around, except in some of the larger cities in the United States, and comparable cities elsewhere, the answer is no, you don't absolutely need an agent. But if you are already established and want all of your work arranged for you by an experienced organization or person, then it is well worth the fee—usually 10 to 15 percent of your earnings.

Agents come in all shapes and sizes. You have to be very careful about them, as some of them will make you think that they are working very hard to place you, while in reality they are merely trying very hard to package their own properties, generally for television, in order to obtain a producer's share of the profits rather than simply an agent's percentage. Some of them know very little about the radio and/or the cable business and will be unlikely to get a job for you in either unless it is as part of a large package.

I'm a legal chauvinist; I believe that a good lawyer who knows what has to be negotiated with a network, station, or production company on behalf of a performer, writer, or producer can do the same job at far less cost than most agents. Hopefully, if the lawyer is also the personal attorney for the client involved, he or she can set up a corporation that may be a good vehicle for the client to operate in, handle the tax aspects of the transaction, and review any contract proposed by the employer or draft one that has all of the necessary terms, without taking a fee out of *all* the client's earnings.

That doesn't mean that there isn't a place and a great need for an agent. There are some very good ones. In some of the larger organizations, many of the agents have attended law school and have the basic knowledge to do what a lawyer might pass on. Lee Stevens, president of the William Morris Agency, one of the largest talent agencies in the world, attended the New York University School of Law with me while he was working at the agency, where he had started in the mail room. Among other sensational deals Lee has worked on was the one in which he talked Barbara Walters into leaving NBC's "Today" show some years ago, and negotiated a long-term, million-dollar contract for her at ABC. He and most of his people are smart, responsible agents, and they have made some marvelous deals for a number of clients in broadcasting, as well as in other areas of the entertainment business. If you can get one of their agents (many of whom also started in the mail room) to represent you, you aren't necessarily on your way to making a lot of money in a telecommunications programming position, but you certainly should have some degree of success if you have the talent.

There are also good agents at International Creative Management, or you might be able to persuade a Don Buchwald or a Susan Smith (who have their own agencies in New York and Los Angeles) to be your agent. They have great rosters of clients.

If you want only to have a lawyer represent you or to package a program that you are about to sell to a network (and you're prepared to pay their fees), you might want to call Leonard Franklin or Michael Rudell at their firm (Mike has also written an excellent book, *Behind the Scenes,* on practicing entertainment law). Or you could refer to Robert Montgomery, at Paul, Weiss, Rifkind, Wharton & Garrison; Melvin Simensky, who also teaches entertainment law at the New York University Law School; or Ronald Konecky. All of them have offices in New York. In California, you might want to contact Mitchell, Silberberg & Krupp in Los Angeles or Rosenfeld, Meyer & Susman in Beverly Hills. I have listed some of the professionals I know in the telecommunications field in Appendix D in this book, with their addresses, to make finding them a little easier for you.

There are certainly many qualified lawyers around the country, but not too many who understand the programming side of the telecommunications business, although a few law schools now offer courses in the field. Most of those are given in New York or Los Angeles, but if you aren't in one of those cities, knowledgeable lawyers might be recommended by your local bar association. In any event, even if you have an agent and/or a business manager, you should have both a lawyer and an accountant review whatever you are doing, especially any agreement that you are asked to sign.

If you are already in the business and achieving some degree of success, it is even more important that you have proper counseling and good business management, because that *is* the way that you make millions in the telecommunications field. You should be receiving the benefit of good tax advice and using every possible technique to preserve as much of your earnings as you can and to increase your assets.

Corporations and Pension Plans—Using Professional Advisers

If you are already in the telecommunications programming business in one aspect or another and earning good money, you should be taking steps to preserve your income and possibly to increase your net worth. Unfortunately, a number of people in the telecommunications industry either ignore the professional assistance that they should have or very often fall into the hands of what I call "unprofessional professionals." I can't do anything about persons who earn hundreds of thousands of dollars a year through their creativity or talent but choose not to have a good accountant and a competent, devoted lawyer counseling them.

I would, however, like to do all that's in my power at least to rid the legal profession of some of the lawyers who try to combine their practice with an accounting practice and offer a "total service" to people in the communications field, including handling their money and investing it, but usually do a poor job in each area. They frequently decrease the client's assets rather than increase them, putting assets into highly speculative enterprises, such as real estate projects in which the "professional" firm itself has a major share interest or into wild tax shelter investments, where years later the government comes after the client to pay tremendous taxes, plus interest and penalties, for recapture of excessive depreciation deductions or other cause.

There are good lawyers in whom you should be able to put your faith, and there are very good accountants who can handle all of your business on a day-to-day basis in a most professional way. They will charge only for the hours they and their staffs put into doing your work, without charging a tremendous percentage fee for doing it and without putting you into ridiculous investments, except perhaps those that you choose yourself. There are also some good investment advisers around, but they generally want to handle large (at least $500,000 of principal) accounts and with full discretion, which you may not want to give.

Over the years I have been able to handle clients, together with a good independent accountant, an investment adviser (one who will work with the client and me, meeting periodically to review all investments), and/or a securities broker—and, when required, a pension plan expert. I prefer that decisions be made by the clients, or if they can't or don't want to be bothered, at least that we keep them constantly advised of what is being done on their behalf. A lot of other lawyers in many places throughout this country and in other parts of the world do the same thing.

Incorporating to Protect Yourself and Your Money

I've found that incorporating is one of the best tools that someone involved in the telecommunications industry can employ. It not only allows for limited liability in almost every state and nation, but it provides for great flexibility when you are considering some things that you may want to do to avoid—*not* evade—taxes. You can put away some of those high earnings that you may be receiving each year so that you will have the money (and maybe a lot more if you invest it wisely) when you are no longer able or do not want to work or when your program is canceled by the network or the station you work for.

I always suggest that the corporation be formed even before you agree to do something, assuming that you are a talent. If you are involved with producing, you will certainly want to use such an entity as soon as you start thinking about your production, so as to keep a good record of your production costs and limit your liabilities for all of your production activities. In that way, as a talent, you can sign a contract with the network, station, or production company employing you and have all of your checks made payable to your corporation.

You may have to individually guarantee your performance of the agreement, but for all intents and purposes you are a corporation and can do all the things (and safeguard against certain liabilities) that any corporate entity can. You can be the president of your corporation and even the sole director in most states if you are the sole shareholder. But you may want to consider having members of your family, friends, or your lawyer as officers, directors, and/or shareholders of your corporation to avoid what are called personal holding company tax problems. On the other hand, your accountant may tell you that, for certain reasons, you shouldn't mind being treated as a personal holding company and you can still have some of the same benefits. In any event, you can decide what kind of salary your corporation should pay you (according to the IRS, it must be a *reasonable* one in the industry in which you work) and more or less determine your own personal annual income.

If your employer wants you badly enough, he or they will agree to make a deal with your corporation for it to supply your services for whatever amount has been agreed upon. They will sign documents with your company, even though they know that they are only really obtaining your services. However, you may be so wanted that they will go even further than that.

When Henry Bushkin represented Johnny Carson in his last negotiations to renew his contract with NBC for doing "The Tonight Show," it was reported that he had Carson Productions, Inc., which employs Mr. Carson, agree to supply his services for the show a certain number of nights each week for about forty weeks a year, in return for which NBC agreed to pay that corporation about $8 million a year. In addition, NBC was reported to have agreed to arrange financing for Carson's film company to produce some films (it coproduced *Terms of Endearment*), and for his television production division to produce some television programs, not all of which were going to be on NBC.

Obviously, NBC needs Johnny Carson very badly because it makes a lot of money from the advertising carried on his very popular program. The ratings that he brings to the network significantly improve their overall reputation and relationship with their affiliated stations. A very advantageous deal was struck that provides Carson with healthy ordinary income and provides significant side benefits that may earn him (and those associated with him) millions of dollars more in profits from other ventures.

Many other performers have been able to strike similar deals over the years, although the Carson transaction is reportedly one of the richest ever in television.

An additional benefit of the use of a corporation is that you are able to set up a pension plan for yourself and any other employees that you may have on your payroll. You can put part of your income aside each year, invest that money further, and ultimately use the principal and income for retirement. The gains that are made on your pension plan investments are taxed as part of the trust, the contributions that you make to the plan are based upon the salary that your corporation pays you and any others, and you can even buy insurance policies that will either pay you monthly benefits after you retire or significant amounts to your loved ones if you die before your retirement date.

About twelve years ago, I urged a young film director client who directs and produces advertising commercials for television to have a pension plan expert set up a plan for the corporation through which he operated his business. The corporation had six employees, including the director and his hard-working and well-organized wife. Over the years, he has made pretty good money in the commercials business, but without the pension plan he would have been required to pay many more dollars to the Internal Revenue Service than he did; he never would have been able to put away and invest the kind of dollars that he has. He has always indicated that he knows very little about business and taxes. But he is one of the most creative film persons I know, so he takes care of the film business, and his wife and I, as co-trustees of the pension trust, have been taking care of what is to be contributed each year by the corporation to the trust and how the funds there should be invested to make them increase. The trust now has over $1 million in principal in it, and the director and his wife, who have the biggest share of the trust under the plan, don't have many financial problems to worry about anymore. Of course, the other employees have benefited as well. He can concentrate on his advertising work and pursue his ambition to direct a feature film, which I'm sure he is going to make someday soon.

Negotiating a Deal

In negotiating employment agreements for creative people in telecommunications, I never hesitate to ask for as much as I think the traffic will bear. On the other hand, some advisers actually have cost their clients their jobs because they don't really know where the line of trying hard for a client stops and good sense begins. You can't be shy in trying to obtain as many benefits as you can for your side, but you have to have some idea of the employer's limits when it comes to giving you what you may demand, however much he or they may need your client.

GET IT IN WRITING

There is no more important piece of advice that I can give to anyone involved in telecommunications than to make every attempt to memorialize any transaction you or your representative may negotiate with anyone or any organization as soon as possible after the agreement is made. This may take the form of a memorandum or a simple letter of agreement signed or even just initialed by all of the parties. Or it may be a full-blown contract that has been drafted and perhaps changed a few times, then finally signed by the parties. Whatever it is, get it in writing!

For many years, the broadcasting business was notorious for putting programs and people on radio and television and then *maybe* signing a contract afterwards. Performers would go on for weeks, sometimes years, without ever having something in writing from their employer. Well, it doesn't and it shouldn't work that way anymore. There is no reason a performer or a director or anyone else employed by a communications organization should not have an agreement in writing to make him or her feel secure, or as secure as one can possibly feel in this business, specifying all the terms agreed upon—especially how one is to be paid. This is particularly true if you are established and you want to deal through your corporation. You can rightfully ask for a contract; you shouldn't accept any excuse for delay in receiving one, even the old line that "our lawyers are working on it."

Most personal service contracts in broadcasting are subject to cancellation in periods of thirteen or twenty-six weeks, so that the network or the station either can correct what it thinks is a mistake in putting on someone the public doesn't like, or retrench if the advertiser cancels after a similar period of sponsorship of a program. For someone who has been appearing for a reasonable period of time on television or radio there really should not be such a cancellation clause in the agreement. For that person a long-term contract should be just that, preferably running three years, with appropriate escalation clauses as to salary and other benefits. This should apply just as much to a news anchorwoman appearing every night for the last two years on a television station in San Antonio, Texas, as for a personality who has been appearing every morning on a New York radio station for many years.

You Can Make a Million, Too

The sum and substance of all this is that there are a lot of people making a lot of money in the telecommunications programming business, and you may be able to as well. You can play a role in making programs by producing them, acting in them, announcing them, moderating them, writing them, photographing or drawing them, or in whatever other role you may want to and be able to play in their development. And you may even make (and keep) a million dollars for doing it.

Get good professional assistance, be certain that you get everything in writing, copyright whatever you can, make sure that your money is invested wisely, and don't be impatient. Now, doesn't that all sound easy? Well, it isn't by a long shot, but it can be done!

What are you waiting for? Get started—even if you've been around for a long time and haven't made it (or kept it) yet. It's never too late to try!

Telecommunications Is for the Support Troops, Too

In every pioneering field of successful endeavor, you'll find a mass of support troops backing up the stars who dazzle out front. Not all of the jobs or all of the opportunities to make a lot of money are visible to the general public. Telecommunications is no exception. Go behind the corporations, limited partnerships, and individuals who are doing so well in this rapidly expanding industry and you'll find a legion of lawyers, accountants, sales representatives, broadcasting and cable system brokers—and many other backup experts—also cashing in on the boom.

I meet or talk on the telephone with many of these people almost every day in my practice and business involvements. But except for the lawyers, I never really knew until recently the kind of money that these support troops were making. Many of them are doing very well, indeed. Whether you want to get involved yourself, or have a son or a daughter who you think should be getting involved, here is some news about those telecommunications support people which you'll find useful.

The Accountants

There are qualified accountants all over the world these days, but very few of them, individually or in firms, have had much experience with the telecommunications business, except for the large accounting firms that act as auditors for some of the outstanding corporations in this industry.

If you are an accountant, or are thinking of becoming one, you might want to consider a position in one of the so-called Big Eight accounting firms, especially one that gets involved with telecommunications companies. Their people get to work on some of the problems of great companies such as AT&T, IBM, GTE, CBS, Viacom, and others I've mentioned. You can get a good education about almost all aspects of a telecommunications business just by doing such work. You also can meet some fascinating people working in the field—stars in the programming area and some very smart and knowledgeable executives and managers as well. And sometimes accountants get to travel on clients' business to exciting places around the world. Clearly, you can make your mark in the firm if you are working with a large or growing company, and the telecommunications clients should fit in that category for some time to come.

Best of all, partnership in such a firm, if you are able to go up a rather structured ladder, may mean a large salary for you, with probably a very good pension in the end. Furthermore, there is always the chance that you will be asked to assume an administrative position in one of the telecommunications companies with which you are working, and that could lead to good stock options that significantly increase your net worth. Former members of accounting firms often turn up these days as chief financial and chief executive officers of leading corporations.

Short of one of the large firms, there are smaller and more specialized firms of certified public accountants involved with auditing corporations in the telecommunications field. They often assist individuals and entities just entering or beginning to become successful. I have recommended clients to some of them, and they have done an outstanding job.

I always thought of accountants as people who pushed around a lot of numbers or prepared tax returns every day. I always supposed that they never made a great deal of money, except in the large firms. However, after finding out more about accountants in the telecommunications field, I have changed my thinking, at least about accountants' earnings, and I've certainly run into many interesting CPAs along the way.

Practicing on your own or as a partner in a small firm can, in the long run, afford you a much greater opportunity to make a lot of money in the telecommunications field than being in a large firm, because you have a much greater opportunity to get in on the ground floor of new businesses that are being formed. Very often an accountant will be asked to take stock in the business as part of his or her fee—or may be asked if he or she wants to invest in it. While this doesn't always pay off, and you have to be careful about becoming too enthusiastic over a client's project, all you need is just one good company that becomes successful to make such an investment really worthwhile.

There's a dynamic accountant in New York, Phil Zimmerman, who doesn't quite fit the popular conception of accountants. He and a few of his partners have been involved in the telecommunications field (including involvement with clients who have been buying and selling radio stations). I like to play tennis with Phil in midtown Manhattan early on weekday mornings, both for the exercise and because it gives me an opportunity to pick his brains about broadcasting acquisitions and operations. He's always bright, cheerful, and cooperative, and I've found he doesn't hesitate to have his firm bill me or my client for his sage advice, even if given before regular working hours.

I haven't had the guts to ask Phil what his net worth is, but I have an idea that he and his firm, Paneth, Haber & Zimmerman, have done quite well, to some degree because of their clients in the telecommunications business. Phil's and the other accounting firms I know of in the telecommunications field are primarily involved in four general areas of communications: advertising, broadcasting, entertainment, and publishing. In broadcasting, they audit and/or advise television and radio station owners, closed circuit television marketing companies, broadcasting brokers, sales representatives, and time buying agencies. They handle the business affairs of some noted actors, actresses, composers, directors, motion picture producers, and musicians. (There's nothing like meeting a few stars every once in a while to liven up a business—and to make it lucrative.) They also do work for television and radio syndication companies, talent agencies, television commercial production companies, theatrical production companies, and writers. Their connections often provide them a chance to take a stock interest and/or invest directly.

It's essential that a CPA understand his client's business before he can even think of being successful. When Phil and other accounting experts in the telecommunications industry started taking on clients in the business, they weren't familiar with the special accounting techniques and technical language used in each area. For example, Phil didn't know what a "log" was. He had no idea about the differences between selling, technical, program, and general and administrative expenses—the different categories of expenses in connection with operating a radio station. He was not aware of the typical ratios for successful radio stations regarding size of market, billing volume, and type of station, all of which are very important in guiding radio station owners in their business operations.

So Phil and some of his partners went to telecommunications industry seminars and meetings, and they read extensively in the field. Phil even attended one of my classes when I had as a guest lecturer Ira Goldstein, a noted broadcasting lawyer (and currently general counsel at Metromedia, Inc.), who discussed the typical asset purchase in radio acquisitions. Phil has gained wide knowledge and now uses it to assist many stellar clients.

Self-study is essential if you want to succeed in the accounting profession, and especially in telecommunications. Through planned reading, you become aware of other sources of technical and accounting information that can help you serve clients involved in telecommunications.

In each of the areas in which these accountants have clients, there are trade associations that gather statistics important to the financial health and success of members of the industry. These associations offer other information that will help CPAs understand their clients' businesses. As an example, the National Association of Broadcasters publishes annual statistics based on a survey of all broadcast properties. This data is very helpful in determining whether or not a particular radio property is doing well or poorly, or even if it should be acquired or not, as compared to similar stations in similar markets. The NAB is located at 1771 N Street, N.W., Washington, D.C. 20036, and you can write them to obtain a list of their current materials and the prices for each.

As Phil Zimmerman said: "Education on industry practices should not be confined to reading in your office or home. It is important to mingle with the people in the industry. This helps not only in learning but in providing a market for an accountant to operate in.

"In the field of radio, for instance, there are periodic seminars held on acquiring radio stations or in improving the operations of radio stations. At these meetings you have the opportunity to meet people who own and operate radio stations, bankers who finance them, attorneys who provide legal advice to them, and brokers who find radio stations for people interested in buying them. When you meet and talk to these people, you find out how you can better relate to them and become more helpful to them in their chosen fields. If you do a good job in helping them, many times they recommend you to others.

"[Service examples include] providing a complete package of financial information to your client when he applies for bank financing in an acquisition, such as gathering prior-year financial statements and arranging them in good order; preparing a projection accompanied by all of the assumptions made by your client for the term of a bank loan or at least five years; devising a feasible income tax savings plan for the station, which you may explain to the bankers; and showing how the loan can be amortized in a way that will meet the requirements of the bank's credit committee. As a result, a banker may recommend you to another potential client.

"To be successful as an accountant in the telecommunications business, you have to bear certain important points in mind. In broadcasting an accountant often finds that he is dealing with smaller companies (and maybe some pretty talented but disorganized people) so that he has to give a lot of administrative assistance as well as provide audited financial statements. This means that those clients need more personal attention from the CPAs than those in organizations that have large administrative staffs. So the accountant has to act as more of a comptroller and financial adviser than he would for a large manufacturing company. People involved with broadcasting often are more sales-oriented than those in manufacturing and trade, and less financially sophisticated, so the accountant must play a bigger role—and probably charge higher fees.

"In the entertainment and programming field, people are generally highly creative; they tend to be more interested in what they are doing than in the financial consequences of what they are creating. They generally have to be treated with kid gloves and a firm hand. Some entertainment clients ask accounting firms to handle the collection of their money and the payment of their bills. The accounting firms mostly charge for these services on an hourly rate, as they do for other professional services. In many cases, the rate per hour is lower than for their professional services, since they can use a less highly trained person to do this work than would be required to do a certified audit. Some CPAs charge a percentage of the gross money that they collect for these clients.

"These CPAs benefit when the client is very successful. On the other hand, you may even lose money when you service a relatively new performer, writer, or other talent. People in the entertainment industry rely on their CPAs to help set up formal or informal budgets, so that when it comes time to pay bills or income taxes there is enough money left over from work previously performed to do so. Also, people in the entertainment field often have a limited professional life during which they earn high incomes."

Phil and others recommend that after training yourself as a specialist CPA, you train a staff specializing in telecommunications accounting. The CPA always has to bear in mind that he is in a service industry providing personal services, and these services must be rendered in a timely manner so that clients can make informed business decisions.

To be a respected and reputable CPA in telecommunications does not require your being in a major city. There are radio and television stations, cable television systems, cellular radio-telephone distributors, marketers of paging devices, and more, in mid-size and even small cities and towns across the country and abroad.

As an example, take the Sconnix Broadcasting Group. It is a partnership (formerly operating as a corporation) that owns a group of radio stations located in cities throughout the United States, and valued at in excess of $25 million. The gross income from their operations is probably over $10 million. Its headquarters are in New Hampshire because the president of the company likes living there and manages one of the stations located there. Last year, the group owner switched auditors from a large national accounting firm to a local New Hampshire firm composed of good, professional people who have taken the time to learn something about the radio industry.

There are a lot of people throughout the world these days who won't make a move without consulting an accountant. The opportunities in the telecommunications field for more and growing clients involved with the new technologies means more and bigger clients for accountants and accounting firms of all sizes in various places, provided that they know something about the industry. I find calculators and computers helpful, but to me calling upon people such as Phil Zimmerman is even better.

The Television and Radio Sales Representatives

Television and radio sales representatives work in a fast-paced, highly competitive atmosphere. Sales requires hard work in the office, as well as skill and confidence in dealing face-to-face with clients in both business and social settings. Sales representatives bring in large revenues for their companies and their clients, and they generally are rewarded with huge salaries and fringe benefits. In 1982, total billings for television stations and networks were $12,152,764,400 and for radio, $4,470,000,000—and in 1984, approximately $14,000,000 for cable. At the larger New York City firms, sales reps can expect to earn a salary of $50,000 to $75,000 after one or two years. Fringe benefits include large expense accounts and opportunities for travel. The work environment provides some of the glamor and excitement of the entertainment industry. Part of the work is wining and dining clients and going to frequent parties. For energetic, people-oriented individuals, a lucrative, exciting career can be had as an advertising sales rep.

The sales rep sells television, radio, or cable television

time to advertisers either directly or through an advertising agency buying department. The sales rep may work for the television or radio station or network directly or for a "rep firm"—an independent company that sells time for stations it has contracted to represent on a commission basis. The networks have in-house firms to represent network-owned and operated stations.

The sales rep is generally assigned to a territory. The rep visits the advertising agency or, in the case of local sales reps, the advertisers directly, to determine what products the advertisers are interested in promoting, what audiences they are interested in reaching, and what kind of "buy" their budget will permit. With this information in hand, the rep compiles the appropriate rating information, usually with the aid of a computer, and includes prices from which negotiations can begin. (Of course, competitors at other stations in the market are doing the same.) The rep must then use this information to sell time to the client by delivering more of the desired audience at the lowest rate.

Advertising time is broken down into "spots" of ten, thirty, and ninety seconds. The spots can be sold individually—e.g., one spot on "Dallas" on July 30th, on WCBS-TV in New York. Or the advertiser may prefer to buy "network" time—the same program on the same date to be seen in every market in the United States where "Dallas" is airing. Of course, network time is more expensive, but it is also more cost efficient for the advertiser looking to enter many markets.

The average thirty-second prime-time spot on network television these days costs $100,000. A Super Bowl minute can run as much as $1 million. A spot on an individual station can range from $10 to $15,000, depending on the market. Radio spots cost from $750 or more in large markets to less than $1 on small-town stations. Cable television rates range from $2 to $400 per thirty-second spot.

The rates for advertising time are set by the station with the help of the sales reps and are negotiable according to supply and demand. The desirability of the spot is determined by the time period "rating." A. C. Nielsen Company and Arbitron Company are the two rating services that survey public viewing patterns and rate programs on a daily basis. A higher rating means more people are watching, and that means higher rates. The system gets further complicated as the ratings are broken down into specific categories, allowing the advertisers to zero in on a particular audience. For example, a BMW automobile dealer may be looking for an audience of affluent men twenty-five to fifty-four years of age. The ratings would probably lead them to advertise on sports, news, and prime-time adventure programs and steer them away from daytime programs, which attract low numbers of male viewers. The sales rep must consider all of this information when negotiating the sale with the client.

The sales rep enjoys a lucrative, exciting, and highly sought-after position. Women and minorities, until recently absent from sales staffs, are now enjoying the success previously reserved for white males. Remember that in every market that has a commercial television or radio station or a cable television system, there must be a sales staff of at least one. Broadcast organizations are constantly looking for good salespeople to go out and pull in the advertising that creates their cash flow.

The most competitive place to begin a career in advertising sales is in the major markets. In those markets, some previous sales or market research experience generally is a prerequisite to an offer from a station or large rep firm. Smaller firms may hire college graduates with business degrees. In addition to experience, employers often require their sales staff to take part in a training program that can last up to six months.

Smaller markets often serve as a training ground for inexperienced salespeople. Offices in major markets often pluck their sales staff from regional branches, so the smaller markets are good places for sales reps to begin. Smaller markets and regional sales offices provide more visibility to ambitious young people than small rep firms or local independent stations do. The networks have sales staffs in offices nationwide, and there are regional sales organizations throughout the country.

Cable television advertising time is sold through local and regional sales reps. Often, sales reps handle all of the advertising sales business for the cable operation. Where a cable operation is represented by one of the larger cable rep firms, the agreement may allow the cable system to handle its own local sales.

Some sales reps never tire of selling, so they continue to do so until retirement. Others prefer to move into management positions. At rep firms, a move to manager is often the first step into management. The group manager oversees a team of salespeople and helps to maintain good relations with clients. From group manager you can move up the ranks of the company and take on responsibilities such as bringing in new clients and overseeing regional offices.

Some reps prefer to leave their firms and sell locally for a station or to work for a client station as station sales manager. Going any of these routes can result in a financially rewarding career.

A good example of how you can make your fortune in television or radio sales is multimillionaire Bob Duffy. At forty-three years of age, he is creating new opportunities to make millions for himself, his company, and his clients.

Bob Duffy, a graduate of Colgate University, was a top draft choice in the National Basketball Association. He played professional basketball for three years, then he was cut from the team. Bob moved back to Colgate to become the youngest major college coach in the history of the NCAA. After two and a half years of coaching, he left Colgate in search of a new career in business.

At twenty-seven, Bob got a job working for Eastman

Radio Reps as a radio sales trainee in their Detroit, Michigan, office. Within two years he was back in New York selling radio time for Eastman. In the next three years Bob moved from salesman to sales manager to head of sales for the company to executive vice-president of Eastman. He attributes his ability to race up the corporate ladder more to perseverance than anything else. Bob advises, "Never accept 'No' for an answer," giving as example the story of how he won his biggest client. The client was located out of town. Both Bob and his competitor made their presentations on Thursday afternoon, and afterwards the competitor flew home for the Labor Day weekend. Bob went back to his hotel room and spent the weekend making phone calls and being available for questions. He won the account and has had it ever since.

In 1973, Bob left Eastman for a cut in salary but a chance to co-run Christal Radio, a rep firm bought by Cox Broadcasting, in an attempt to save it from bankruptcy. He spent five years turning Christal around, then in 1978 he decided to move on to set up his own firm. However, Cox offered to sell Bob the company for approximately $30,000 and five years of sweat equity—Bob to earn 20 percent of Christal each year. Five years later, in March of 1984, Bob sold Christal Radio for a figure in the range of $18 million to Katz Communications, Inc.

Besides running Christal Radio, Bob Duffy created a new company—Duffy Broadcasting—through which he has acquired four radio stations. And he intends to buy more. Though the initial investment in Duffy Broadcasting was $6 million, Bob only invested $120,000 of his own money, convincing venture capitalists to put up the rest, using what he calls "positive persuasion through positive persistence."

Bob's hard work and positive, confident attitude have made him a success. "Inexperience," he says "should not be an impediment to new opportunities. When a potential employer rejects you due to lack of experience, go sit in the reception area and come back every morning at 8 A.M. to greet the top people before they disappear into their offices."

Bob Duffy believes that his sports training has enabled him to feel challenged rather than pressured by competition. He also believes that anyone who can compel people to listen and then works hard once he or she opens that door can achieve the same success.

As Bob Duffy has indicated by example, being a sales representative in the telecommunications business can provide a very rewarding career. Clearly, as the industry grows the opportunities increase with it.

The Brokers

With more people getting into ownership of broadcasting and cablecasting properties, the business of putting together sellers and buyers has become increasingly lucrative. As television stations are being bought and sold for figures in excess of $100 million, and even small radio station prices go over $1 million, you can imagine the commissions being earned by the brokers, who may take anywhere from 1 to 8 percent of the selling price in each transaction.

Most of the brokers are located in and around Washington, D.C., although some have supplementary offices in large cities such as Chicago, Dallas, San Francisco, and Los Angeles. With the growth of cable television, Denver, Colorado, has also become a base for certain brokers. Washington has been the historical base because a number of brokers once worked at the Federal Communications Commission or were involved with group owners of broadcasting properties that had their headquarters there. I have often wondered why there are not more brokers operating out of New York City, where so many communications companies are located. Perhaps it is because so many people in New York would rather be owners and bankers than brokers.

As the rules for ownership change and as the telecommunications industry evolves, there will be a tremendous growth in this area of the industry and an opportunity for many more people to become involved.

I don't know of any women in the broadcasting brokerage business, but I also don't know of any reason there should not be many entering the field. Just as women have become significantly involved in the real estate brokerage business, and some have made a lot of money at it, I suspect that women soon will be coming into broadcasting brokerage, too, and they will probably do very well at it.

The largest brokerage firm in the telecommunications business today is probably Blackburn & Company, Inc., headed by Jim Blackburn, son of the founder. Based in Washington, D.C., Blackburn also has offices in Chicago, Atlanta, and Beverly Hills, California. One of its leading brokers, Joe Sitrick, seems to be involved in some way in almost every large transaction these days. The company's spokesmen seem to be present at every national and state convention of broadcasters and cable system people, and they are always striking deals that seem to be as good for the buyers as for the sellers.

I happen to like Cecil L. "Lud" Richards, who has pretty much operated on his own out of Falls Church, Virginia, but now also has an office in Northbrook, Illinois. Lud Richards is always forthright and zealous in representing the seller in a transaction, and he also tries to be fair and helpful to the buyer. Lud was the broker in the first transaction in which I was an equity participant. He certainly was very professional and helpful to both sides in that deal, which has so far turned out to be a good one for us.

Gammon & Ninowski is a rather new firm of media brokers with offices in Washington, D.C., Los Angeles, San

Diego, and Jackson Hole, Wyoming. They specialize in medium and major market radio stations, television stations, and satellite transponders, and they seem to be approaching the business in a very professional manner. They are even willing to represent a buyer for only a flat fee, which is rather unusual these days.

Some brokers operate only on a regional basis, which may afford them a better opportunity to satisfy clients and obtain commissions through concentration on properties in areas that they know well. An example of this type of broker is Bob Kimel's New England Media, Inc. Bob is a tried and true New Englander. He really knows the territory and has first-hand knowledge of operating radio stations in small markets in the New England states. He has brought together sellers and buyers in those states, sometimes people from as far away as California, who wanted to enjoy the beauties of, say, Vermont and decided to own and operate a radio station in a small town with hopefully satisfying economics and a rewarding way of life.

The Ted Hepburn Company, based in Cincinnati, Ohio, is probably *the* acquisition specialist in the cable business. It has been involved in some very big deals of late.

Almost all of the media brokers are listed each year in the *Broadcasting/Cablecasting Yearbook*. Unless you feel qualified to enter this field immediately, I suggest that you start by going to work for one of the established firms to get the experience and contacts you need to succeed in this field. Most firms are on the lookout for more and better people, so you may do quite well sooner than you expect.

The Lawyers

Practicing law and specializing to some degree in the communications field has afforded me an opportunity not only to see my firm receive good fees from clients in the field, but also to become directly involved in equity participation in organizations that may very well increase my family's net worth someday. It has provided a very satisfying career for me and has allowed me to teach and write. I recommend it highly to anyone interested in the profession or now involved in any way.

I came into the field through working in a growing law firm in New York that had outstanding senior partners and some very interesting and diverse clients. Originally, I worked with clients in, among other things, the film, theater, and advertising fields but had very little to do with the telecommunications industry. As I obtained some of my own clients, especially one that was involved with computer-assisted learning systems (teaching machines), I learned more about telecommunications technologies and business. One of my partners, Orrin G. Judd, through being the leader of the Protestant Council in New York, became involved with the citizens' group movement in broadcasting, and another partner, Earle K. Moore, took on the representation of the Office of Communication of the United Church of Christ.

In the mid-1960s, "Dick" Moore conducted one of the leading license renewal proceedings in the United States, involving WLBT-TV in Jackson, Mississippi. He then became counsel in a number of other matters involving citizens' groups and radio and television stations. When we formed a new law firm in 1968, I took on a case involving a one-kilowatt radio station (KAYE) licensed in Puyallup, Washington, near Tacoma. It had some very offensive programming—anti-black, anti-Semitic, anti-almost everything. I learned everything I could about broadcasting law and regulations and ended up spending a month in Washington trying to have the station's license renewal application denied by the FCC. The case went on for almost another seven years without a final resolution, until the owner finally was permitted to sell the station. I stayed involved for only about two of those years, but I learned a great deal about practicing in the communications field.

From there, I became involved with purchases of broadcasting properties, representing a group involved in the early communications satellite proceedings at the FCC, and in establishing satellite network programming companies. I also consulted on cable television and videocassette matters. I've had a diverse group of clients from whom I have learned much about the industry.

There are literally thousands of men and women with legal training involved in the telecommunications field in one way or another, and I'm sure that a lot of them are or will be doing quite well, not only as lawyers but also as business people. The opportunities are simply tremendous, perhaps so vast as to create great difficulty in making a choice on where to enter or where to go next. If you or someone you know is interested in getting into the field, don't hesitate. Get into it!

While many of the law firms that are called specialists in communications law are in Washington, D.C., there are a number of practitioners and firms involved with clients in cities throughout the United States and, of course, the world.

Some experience in a governmental agency, such as the Federal Communications Commission or the Federal Trade Commission, even the State Department, will certainly help an aspiring communications lawyer to cope with the tasks that he or she may become involved with. It will also offer very valuable contacts with law firms and corporations concerned with telecommunications—valuable now and when you decide to leave government and earn some real money.

All the large and even some medium-sized telecommunications corporations have hordes of lawyers working for them. Before the breakup of AT&T, it had over 450 lawyers on its payroll. Most of them, plus many new additions, are now either still with AT&T or with one of the regional holding companies established through Judge Greene's orders. MCI Communications Corporation now has at least

15 lawyers on its in-house staff and still retains a number of outside law firms to handle its enormous legal problems and litigation. The television and radio networks and many of the independent broadcast organizations have growing in-house law firms, as well as outside counsel. They are constantly looking for people with capability not only to advise on legal matters, but to be directors of business affairs—to make the deals for the programs to be aired. Cable television companies, such as HBO, Warner-Amex, and Viacom, when adding lawyers to their staffs, make it a policy only to hire those with about three years of law firm experience, preferably in the corporate area, although there are occasional exceptions to this rule.

While I obviously am prejudiced, I always have advised my students that getting into telecommunications in some way will prove professionally satisfying and probably monetarily rewarding in the long run. If you have any interest in the industry, you had better try entering it in any way that you can as soon as possible. Good luck in your efforts, and remember to use persistence, patience, and fortitude!

7

Some More Thoughts About the Future

There are a number of people in this world who can intelligently review what has happened in a certain field, understand what is presently occurring, taking into account what is fact as against what's simply good public relations, and make reasonably accurate prognostications about the long-term future. There are many people who do that in business every day, and often they stake their own money on what they predict will happen, then reap the benefits in the future.

Much of what I've said here is based upon what I believe will occur in the field of telecommunications for the rest of this decade. I'd like to believe that there will be life on this planet (and perhaps others) after 1990 and, since I have the benefit of this forum without charge, I'd like briefly to indicate what I think will be the state of telecommunications by the year 2000.

I'm not sure that I know what the value of the dollar or any other currency will be by that time, so it may not be meaningful to say that you can make a million by doing this or that in the industry between now and the end of the century. But it does seem to me that if you are able to visualize the business situation fifteen years from now and do something about it now, you can do quite well by then.

It seems almost certain that by 2000 there will be a much greater use of telecommunications technology throughout society, bringing people everywhere closer together. This will make it almost mandatory for people of varied cultures to learn what others are doing. I can't say that it will be a better world, but it definitely will be more interdependent.

By the year 2000, there will be full construction of cable television systems not just in the United States, but in most of the industrialized countries of the world as well. There will be more extensive use of multiple channels of programming offered to subscribers and full utilization of cable systems for all supplementary services. There will also be full employment of communications satellites to permit instant relay of programming from and to anywhere in the world—and extraterrestrially, too. Where cable can't penetrate because of physical obstructions and economic limitations, homes and offices will have very small receiving dishes on roofs or even at a window to receive programming of all forms on direct broadcast satellite (DBS) technology. However, I don't expect that DBS will be a large business.

The full utilization of videodisc technology will emerge some time before the end of the century; the laser disc will foster interactivity for educational and informational purposes, shopping, entertainment, and any number of other uses. However, it will no longer use an expensive and cumbersome disc; instead it will employ a digitally produced card the size and texture of an American Express card, which you will be able to carry around in your pocket. There will be as much information on one card as may presently be incorporated in an entire printed set of the largest encyclopedia. The disc player will permit random access and instant review, so that you will be able to find any "page" of information that you want instantaneously. It remains to be seen what use such amazing technology will be put to, but hopefully some good research and development will occur by then so that universally useful programming is produced. Perhaps this will not completely replace the videocassette recorder by the year 2000, but the business of selling and renting prerecorded videocassettes will be affected by the ability to make and market both low-cost prerecorded "videocards" and the technology to play them.

Radio and television stations, satellites, radio-telephones, and paging devices will still, because so many people and frequencies will be involved, require governmental licensing, essentially for engineering purposes. The spectrum will not get too much larger, but all of the regulations that have been so cumbersome for operators of broadcasting facilities will be generally gone. Ideally, those who put on programming will be a bit more responsible than they have been up to now, and we will see greater attention paid to meaningful as well as entertaining programs, particularly for children. I still have hope in this area, and with a new generation of people taking command of programming on the technology employed, this should come about.

I believe that our homes will focus around a *telecommunications center* containing the more sophisticated versions of computers, telephones, fiber optic cables, satellite hook-ups, copying machines, stereo, high density television, and good old-fashioned radio, to operate our dwellings and to furnish us with information and entertainment in various forms. This center may also serve as an office for those who must or simply wish to participate in commercial activities without traveling to an office or store. Architects already are contemplating how these centers will be structured. For better or worse, they will be the central feature of the new century's establishments.

I am certain that within fifteen years only a few large organizations will control much of the electronic media in the United States. While this should make a lot of people who have owned and operated successful radio and television stations a lot wealthier—I don't want to guess what prices such properties will command by then—I hope that our children and grandchildren will not be adversely affected by such conglomeration. Newspapers will still be read, but many of the organizations that control the printed press will also control a good deal of electronic media. Does this mean that there will be the kind of mind control that so many fear will result from possible multiple and cross-ownership in this country? I doubt it, because people in all these organizations will be trying to operate their businesses to maximize profitability, so they will try to appeal to what they perceive as public need and desire.

I'm much more fearful of the matter of invasion of privacy in the use of the new technology; it doesn't appear certain at all that protection of privacy can be insured to the degree that we hope for. I pray that we all will challenge intrusions on our privacy wherever and whenever they may or do occur. If we are to have a viable society, we will have to use extraordinary efforts to maintain the fundamental basis of our democracy, the principles of the First Amendment and the right of privacy.

Not everyone will have a telephone on the wrist and in the car by the year 2000, but certainly the means of communicating voice and data to and from almost everyone in every place in the world will be available and employed by all working societies. Messages will be relayed over computer terminals, and facsimile machines will allow for hard copy to be sent to almost everyone and kept in that form if needed.

I'm not so sure that all of us will be traveling through space by 2001, but obviously we won't have to travel as much as a lot of people do these days. The use of teleconferencing setups will be extensive; universities and medical centers will be sharing their knowledge and their leading professors by using telecommunications facilities daily.

The world in the year 2000 *will* be a different one. It will be filled with even more bright, creative people able to use this new technology, not simply for personal aggrandizement, but hopefully to make the world a better place to be in. I'm looking forward to still being around at that time. It should be extraordinary and a lot of fun to participate in.

Appendix A
Some Telecommunications Publications

Magazines

American Telecom Quarterly—American Telecom, Inc., 3190 Mira Loma Avenue, Anaheim, CA 92806. Quarterly. Free.

Broadcast Banking—Paul Kagan Associates, Inc., 126 Clock Tower Place, Carmel, CA 93923. Bimonthly. $395 for twenty-four issues.

Broadcasting—Broadcasting Publications, Inc., 1735 DeSales Street, N.W., Washington, DC 20036. Weekly. One year $60, two years $115, three years $160. Single issue $2. Annually publishes *Broadcasting/Cablecasting Yearbook,* $80.

Broadcast Stats—Paul Kagan Associates, Inc., 126 Clock Tower Place, Carmel, CA 93923. Twenty-four issues per year $325.

CableNews—Phillips Publishing, Inc., 7315 Wisconsin Avenue, Suite 1200N, Bethesda, MD 20814. Fifty weeks $227.

Cable TV Investor—Paul Kagan Associates, Inc., 126 Clock Tower Place, Carmel, CA 93923. Bimonthly. $5.95 for twenty-four issues.

Cable TV Law Reporter—Paul Kagan Associates, Inc., 126 Clock Tower Place, Carmel, CA 93923-8734. Twenty-four issues $349 annually.

CableVision—International Thomson Communications Inc., P.O. Box 5208 T.A., Denver, CO 80217-85208. Weekly. One year $64, two years $115.

Common Carrier Week—Television Digest, Inc., 1836 Jefferson Place, N.W., Washington, DC 20036. Weekly. One year $327.

Communications News—124 South First Street, Geneva, IL 60134. Monthly. One year $25, two years $40.

Communications Week—CMP Publications, Inc., 111 East Shore Road, Manhasset, NY 11030. Biweekly. One year $65, two years $120.

DataCable News—TeleStrategies, Inc., P.O. Box 1218, McLean, VA 22101, or call (703) 734-7050. One year $287.

E-ITV—C. S. Tepfer Publishing Company, Inc., 51 Sugar Hollow Road, Danbury, CT 06810. Monthly. One year $15, two years $28, three years $40.

Electronic Media—Crain Communications, Inc., 740 Rush Street, Chicago, IL 60611. One year $25.

Finding Telecommunications Information—Phillips Publishing, Inc., Suite 1200N, 7315 Wisconsin Avenue, Bethesda, MD 20814. $595.

1984 Interconnect Survey—Phillips Publishing, Inc., Suite 1200N, 7315 Wisconsin Avenue, Bethesda, MD 20814. One year $127.

RadioActive—National Association of Broadcasters, 1771 N Street, N.W., Washington, DC 20036. Monthly. $26 per year, included in membership dues.

Radio & Records—Radio & Records, Inc., 1930 Century Park West, Los Angeles, CA 90067. Weekly. $215 per year, or $60 per quarter.

RadioNews—Phillips Publishing, Inc., 7315 Wisconsin Avenue, Suite 1200N, Bethesda, MD 20814. Biweekly. $197 per year.

Radio Only—Inside Radio, Inc., Executive News, 1930 East Marlton Pike, Suite S-93, Cherry Hill, NJ 08003-4210. Monthly. One year $25.

State Telephone Regulation Report—Telecom Publishing Group, Capitol Publications, Inc., 1300 North 17th Street, Arlington, VA 22209. Biweekly. Price uncertain.

Telecommunications Alert—Telecommunications Alert, One Park Avenue, New York, NY 10016. Monthly. Seven months $89, one year $149, two years $275.

Telematics—Law & Business, Inc., 855 Valley Road, Clifton, NJ 07013. Monthly. One year $160.

Television Digest—Television Digest, Inc., 1836 Jefferson Place, N.W., Washington, DC 20036. Weekly. $406 annual subscription, plus $20 for postage and handling. Includes *Consumer*

Electronics. Annually publishes *Cable & Station Coverage Atlas,* $85.

Television/Radio Age—Television Editorial Corporation, 1270 Avenue of the Americas, New York, NY 10020. Biweekly. Subscription $40 year.

Variety—Variety, Inc., 154 West 46th Street, New York, NY 10036. Weekly. Newsstand price $1.50 per issue. Annual $75.

Venture—Venture Magazine, Inc., 35 West 45th Street, New York, NY 10036. Monthly. One year $18, two years $32, three years $44.

Books and Other Publications

Accounting Manual for Radio Stations—National Association of Broadcasters. 1771 N Street, N.W., Washington, DC 20036. $5.00.

All About Cable—Legal and Business Aspects of Cable and Pay Television (Morton I. Hamburg, assisted by Thomas H. Nathan)—Law Journal Seminars-Press, 111 Eighth Avenue, New York, NY 10011. $70 with current update. Current update alone $35.

American Radio—published in spring and fall each year (James H. Duncan, Jr.), Box 2966, Kalamazoo, MI 49003; (616) 342-1356. $58.

Broadcast Lending—A Lender's Guide to the Radio Industry (Robin B. Martin)—National Association of Broadcasters, 1771 N Street, N.W., Washington, DC 20036. Write for current price.

Buying or Building a Broadcast Station (Erwin Krasnow)—National Association of Broadcasters, 1771 N Street, N.W., Washington, DC 20036. $20 members, $60 nonmembers.

Cable TV Financial Databook—Paul Kagan Associates, Inc., 126 Clock Tower Place, Carmel, CA 93923. $60. Two or more $50 each.

Costs of Cable Television Franchise Requirements (William B. Shaw; prepared for The National Cable Television Association; February 1984)—National Economic Research Associates, Inc., Consulting Economists, 123 Main Street, White Plains, NY 10601.

Current Trends in the Direct Broadcast Satellite Industry (Mark

Oderman, Center for Space Policy Inc.)—Phillips Publishing, Inc., Suite 1200N, 7315 Wisconsin Avenue, Bethesda, MD 20814.

Drexel Burnham Lambert Incorporated Research Reports—Office of John Reidy, Drexel Burnham Lambert, Inc., 60 Broad Street, New York, NY 10004. Various reports on industry and various public companies. Generally free on request.

Duncan's Radio Market Guide, 1984 Edition (James H. Duncan, Jr.)—Duncan Media Enterprises, Box 2966, Kalamazoo, MI 49003; (616) 342-1356. $125.

How to Write a Business Plan—American Management Associations Extension Institute, 135 West 50th Street, New York, NY 10020. $89.95 (AMA members $79.95).

Investing in Television (83/84)—Broadcasting Publications, Inc., 1735 DeSales Street, N.W., Washington, DC 20035. $250.

NAB Legal Guide to FCC Broadcast Regulations—National Association of Broadcasters, 1771 N Street, N.W., Washington, DC 20036. $95.

N/E/R/A Quarterly Review—National Economics Research Associates, Inc., 123 Main Street, White Plains, NY 10601. Free.

1984 Broadcast Financial Record—Paul Kagan Associates, Inc., 26386 Carmel Rancho Lane, Carmel, CA 93923. $225.

1984 Bypass Market Analysis and Demand Forecast—TeleStrategies, Inc., 6842 Elm Street, Suite 102, McLean, VA 22101. $1495.

The 1985 Satellite Directory—Phillips Publishing, Inc., Suite 1200N, 7315 Wisconsin Avenue, Bethesda, MD 20814. $197, plus $8.50 handling charge.

1984 Satellite Transponder Supply/Demand Analysis and Directory—TeleStrategies, Inc., 6842 Elm Street, Suite 102, McLean, VA 22101. $995.

Pike & Fischer Radio Regulation—Pike & Fischer, Inc., 4550 Montgomery Avenue, Bethesda, MD 20814. Contains full text of rules, regulations and decisions of the Federal Communications Commission and its Review Board, Bureaus and Administrative Law Judges; pertinent decisions of state and federal courts; treaties; proposed rules and regulations; and FCC forms. "Full service" is $1835 for the first year. Less for only certain portions of full service and for renewals. Looseleaf service, with weekly updates available as part of full service.

Primer on Radio Station Investment (1984 edition)—PK Services Corporation, 126 Clock Tower Place, Carmel, CA 93923. 70 pages. $225.

Primer on TV Station Investment (first edition)—PK Services Corporation, 126 Clock Tower Place, Carmel, CA 93923. $195.

SMATV—A Changing Environment for Private Cable (Dr. Herbert H. Howard and Dr. Sidney L. Carroll)—National Association of Broadcasters, 1771 N Street, N.W., Washington, DC 20036.

Telecom Library, Inc., 12 West 21st Street, New York, NY 10010. Various publications on telecommunications. Send for catalog.

Telecommunications Policy for the 1980's: The Transition to Competition (The Washington Program of The Annenberg Schools of Communications)—Prentice-Hall, Inc., Information Services Division, P.O. Box 511, West Nyack, NY 10995. $125.

Venture Capital and Small Business Financings (Robert J. Haft)—Clark, Boardman Company, Ltd., 435 Hudson Street, New York, NY 10014. $85.

Top Executives in Telecommunications

ABC, Inc.
1330 Avenue of the Americas
New York, NY 10019
(212) 887-7777

Leonard H. Goldenson, Chairman of the Board and Chief Executive Officer; Frederick Pierce, President and Chief Operating Officer; Peter Cusak, Vice-President, Human Resources Management

Advanced Systems, Inc.
155 East Algonquin Road
Arlington Heights, IL 60005
(312) 981-1500

Arthur H. Stromberg, Chairman of the Board; William R. Roach, President & Chief Executive Officer; John A. Kirkham, Senior Vice-President; M. H. Erickson, Director of Personnel

A. H. Belo Corporation
Community Center
Young & Houston Streets
Dallas, TX 75265
(214) 745-8730

J. M. Moroney, Jr., Chairman & President; N. Norvell, Vice-President & Secretary-Treasurer

American Telephone & Telegraph Corporation (AT&T)
550 Madison Avenue
New York, NY 10022
(212) 605-5500

Charles L. Brown, Chairman of the Board; James Olson, Vice-Chairman; Lawrence Conterno, Director of Personnel

John Blair & Company
1290 Avenue of the Americas
New York, NY 10104
(212) 603-5000

Jack W. Fritz, President & Chief Executive Officer; Harry B. Smart, Vice-President, Chairman & Chief Executive Officer—Blair Television Division; John N. Boden, President—Blair Radio; Mary Jane Hurley, Director of Personnel

Capital Cities Communications, Inc.
24 East 51st Street
New York, NY 10022
(212) 421-9595

Thomas S. Murphy, Chairman of the Board; Daniel B. Burke, President; Joseph P. Dougherty, Executive Vice-President & President—Broadcasting Division; William R. James, Executive Vice-President & President—Cable Television Division; Sharon Roberts, Personnel Recruiter

CBS, Inc.
51 West 52nd Street
New York, NY 10019
(212) 975-4321

Thomas H. Wyman, Chairman & Chief Executive Officer; Gene F. Jankowski, Vice-President (President—CBS/Broadcasting Corp.); Richard Plante, Director of Personnel

Chris Craft Industries, Inc.
600 Madison Avenue
New York, NY 10022
(212) 421-0200

Herbert J. Siegel, Chairman & President; Lawrence R. Barnett, Executive Vice-President; Evan Thompson, President—Broadcast Division; Barbara Pappas, Office Manager (Director of Personnel)

Chyron Corporation
265 Spagnoli Road
Melville, NY 11747
(516) 694-7137

Alfred O. P. Leubert, Chairman of the Board; Leon Weissman, Vice-Chairman; Joseph L. Scheuer, President; John A. Poserina, Vice-President & Treasurer; Paul J. Rozzini, Vice-President; Cathy Anderson, Personnel Manager

Communications Satellite Corporation (COMSAT)
950 L'Enfant Plaza, S.W.
Washington, DC 20024
(202) 863-6000

Joseph V. Charyk, Chairman & Chief Executive Officer; Irving Goldstein, President; Robert W. Baumann, Vice-President, Human Resources & Organization Development

Cox Communications, Inc.
Box 105357
Atlanta, GA 30348
(404) 843-5000

Clifford Kirtland, Chairman of the Board; William A. Schwartz, President & Chief Executive Officer; Cynthia Williams, Director of Personnel

The Dow Jones Company
22 Cortlandt Street
New York, NY 10007
(212) 285-5000

Warren H. Phillips, Chairman of the Board & Chief Executive Officer; Ray Shaw, President; Charlene Watler, Director of Personnel

General Electric Company
3135 Easton Turnpike
Fairfield, CT 06431
(203) 373-2211

John F. Welch, Jr., Chairman; Edward Hood, Vice-Chairman; John Burlingame, Vice-Chairman; Frank Oppedisano, Manager, Headquarters Employee Relations

GT&E Corporation
1 Stamford Farm
Stamford, CT 06904
(203) 965-2000

Theodore F. Brophy, Chairman of the Board & Chief Executive Officer; Bruce Carswell, Director of Personnel

Gulf & Western, Inc.
1 Gulf & Western Plaza
New York, NY 10023
(212) 333-7000

Martin S. Davis, Chairman of the Board & Chief Executive Officer; Michael Hope, Vice-President, Finance; Diane Kenny, Director of Personnel

Harris Corporation
1025 West NASA Boulevard
Melbourne, FL 32919
(305) 727-9100

Joseph A. Boyd, Chairman & Chief Executive Officer; John T. Hartley, President & Chief Operating Officer; Bill Maier, Personnel Director

Insilco Corporation
1000 Research Parkway
Meriden, CT 06450
(203) 634-2000

Durand Blatz, Chairman of the Board; Donald J. Harper, President; Richard T. Walsh, Director of Personnel

International Business Machines Corporation (IBM)
Old Orchard Road
Armonk, NY 10504
(914) 765-1900

John R. Opel, Chairman; John F. Akers, President; Barbara
Shimer, Director of Personnel

Katz Communications, Inc.
1 Dag Hammarskjold Plaza
New York, NY 10017
(212) 572-5500

James L. Greenwald, President; Richard D. Mendelson, Executive Vice-President & Chief Operations Officer; Nancy Williams,
Director of Personnel

LIN Broadcasting Corporation
1370 Avenue of the Americas
New York, NY 10019
(212) 765-1902

Donald A. Pels, President/Chairman; E. Blake Byrne, Vice-President—Television; Richard P. Verne, Group Vice-President—Television; Frank Ryniker, Office Manager

Lorimar Productions
3970 Overland Avenue
Culver City, CA 90230
(213) 202-2000

Merv Adelson, Chairman of the Board; Lee Rich, President;
Lucy Svimonoff, Director of Personnel

Malrite Communications Group, Inc.
1200 Statler Office Tower
Cleveland, OH 44115
(216) 781-3010

Milton Matz, Chairman and Chief Executive Officer; Carl E.
Hirsch, President; Gil Rosenwald, Executive Vice-President/Director of Radio; John C. Chaffee, Senior Vice-President/Director
of Television; Elizabeth Nardi, Director of Personnel

Matsushita Electrical Corporation of America
U. S. Headquarters
One Panasonic Way
Secaucus, NJ 07094
(201) 348-7705

M. Matsushita, Chairman & Chief Executive Officer; T. Yamashita, President; Ira Pearlman, Director of Personnel

MCI Communications Corporation
1133 19th Street, N.W.
Washington, DC 20036
(202) 872-1600

William G. McGowan, Chairman & Chief Executive Officer; V. Orville Wright, President & Chief Operating Officer; Kenneth A. Cox, Senior Vice-President; Jerry Inglert, Director of Employee Relations

MMT Sales, Inc.
630 Third Avenue
New York, NY 10017
(212) 599-0899

Gary Scollard, Chairman; Jack Oken, President; Neil Kennedy, Executive Vice-President; Dolores White, Director of Personnel

Motorola, Inc.
Motorola Center
1303 East Algonquin Road
Schaumburg, IL 60196
(312) 397-5000

Robert W. Galvin, Chairman & Chief Executive Officer; William J. Weisz, Vice-Chairman & Chief Operating Officer; John F. Mitchell, President & Assistant Chief Operating Officer; Levy Katzir, Senior Vice-President & General Manager—New Entertainment; Robert N. Swift, Executive Vice-President & Director of Personnel

NCR Corporation
1700 South Patterson Boulevard
Dayton, OH 45479
(513) 445-5000

Charles Exley, Jr., Chairman & President; Steven N. Bowen, Vice-President, Communications; B. Lyle Shafer, Vice-President, Human Resources

The New York Times Company
229 West 43rd Street
New York, NY 10036
(212) 556-1234

Arthur Ochs Sulzberger, Chairman & Publisher; Walter Mattson, President & Chief Executive Officer; Sydney Gruson, Vice-Chairman; William O'Fallon, Personnel Director

North American Philips Corporation
100 East 42nd Street
New York, NY 10017
(212) 697-3600

Pieter Vink, Chairman; Cees Bruynes, President & Chief Executive Officer; Richard Feeney, Director of Personnel

NYNEX Corporation
335 Madison Avenue
New York, NY 10017
(212) 370-7400

D. C. Staley, Chairman & Chief Executive Officer; William G. Burns, Executive Vice-President & Chief Financial Officer; E. G. Guillet, Director of Personnel

PageAmerica Group, Inc.
228 East 45th Street
New York, NY 10017
(212) 286-8901

David Post, Chairman of the Board; Steven Sinn, President; Maryann Eardley, Director of Personnel

Philips Information Systems
4040 McEwen
Dallas, TX 75234
(800) 527-0204

Darrel Baldwin, President; Tony Levecchio, Vice-President, Finance; Sheila Redmon, Director of Personnel

The Plessey Company plc
2-60 Vicarage Lane
Ilford, Essex 1G1 4AQ
England

Sir John Clark, Chairman of the Board & Chief Executive Officer

United States Headquarters:
Plessey Dynamics Corporation
277 Park Avenue
New York, NY 10172
(212) 349-2252

Warren J. Sinsheimer, Chairman of the Board; Joseph Palazzi, Vice-President—Finance; Frank Gillen, Vice-President—Corporate Services

Prentice-Hall, Inc.
Route 9-W
Englewood Cliffs, NJ 07652
(201) 592-2000

Frank J. Dunnigan, Chairman of the Board; Donald Schaefer, President; Donald Caldwell, Director of Personnel

Price Communications Corporation
45 Rockefeller Plaza
New York, NY 10020
(212) 582-1610

Robert Price, President, Chief Executive Officer & Treasurer; Frank D. Osborn, Senior Vice-President; Ellen S. Fader, Vice-President & Secretary

Raytheon Company
141 Spring Street
Lexington, MA 02173
(800) 225-9874
(617) 862-6600

Thomas L. Phillips, Chairman of the Board & Chief Executive
Officer; D. Brainerd Holmes, President; James Hagan, Employment Manager

RCA Corporation
30 Rockefeller Plaza
New York, NY 10020
(212) 621-6000

T. F. Bradshaw, Chairman of the Board & Chief Executive
Officer; R. R. Frederick, President; E. L. Scanlon, Director of
Personnel

Reeves Communications Corporation
605 Third Avenue
New York, NY 10158
(212) 573-8600

Marvin Green, Jr., Chairman of the Board, President & Chief
Executive Officer; William Wetzel, Executive Vice-President;
Ruth Frank, Director of Personnel

Rogers Cablesystems, Inc.
Suite 2602
Commercial Union Tower
Toronto, Ontario
Canada M5K 1J5
(416) 864-2373

John W. Graham, Chairman of the Board; Edward S. Rogers,
Vice-Chairman & Chief Executive Officer; Colin D. Watson,
President

Scientific-Atlanta, Inc.
1 Technology Parkway
Box 105600
Atlanta, GA 30348
(404) 441-4000

Sidney Topol, Chairman of the Board & Chief Executive Officer; John Levergood, President; John Tekowisez, Vice-President, Human Resources

SFN Companies, Inc.
1900 East Lake Avenue
Glenview, IL 60025
(312) 998-5800

John R. Purcell, Chairman & President; L. Steven Minkel, Senior Vice-President—Finance & Treasurer; Richard D. Roberts, Senior Vice-President—Administration & Development

SONY Corporation
7-35 Kitashinagawa, 6-chome
Shinagawa-ku, Tokyo 141
Japan

Akio Morita, Chairman of the Board & Chief Executive Officer; Norio Ohga, President

United States Headquarters:
9 West 57th Street
New York, NY 10019
(212) 371-5800

SONY Corporation of America
Sony Drive
Parkridge, NJ 07656
(201) 930-1000

Norio Ohga, Chairman of the Board; Kenji Tamiya, President; John Stern, Senior Vice-President Human Resources

Storer Television Sales, Inc.
800 Third Avenue
New York, NY 10022
(212) 935-6000

Frances P. Barron, President; Peter E. Murray, Executive Vice-President & General Sales Manager; Laurie Lotter, Director of Personnel

Taft Broadcasting Company
1718 Young Street
Cincinnati, OH 45210
(513) 721-1414

Charles S. Mechem, Jr., Chairman & Chief Executive Officer; David S. Ingalls, Vice-Chairman; Dudley S. Taft, President & Chief Operating Officer; R. D. Gugnon, Executive Vice-President—Television; Carl J. Wagner, Executive Vice-President—Radio & Cable; Don Urban, Director of Personnel

Tektronix, Inc.
Box 500
4900 S.W. Griffith Drive
Beaverton, OR 97077
(503) 627-7111

Earl Wantland, President & Chief Executive Officer; W. Velsink, Vice-President—Research; Les Stevens, Vice-President—Finance; William Walker, Vice-President—Engineering

Tele-Communications, Inc.
54 Denver Technological Center
Call Box 22595
Wellshire Station
Denver, CO 80222
(303) 771-8200

Robert Magness, Chairman of the Board; Dr. J. C. Malone, President & Chief Operating Officer; J. C. Sparkman, Executive Vice-President

TelePictures Corporation
475 Park Avenue South
New York, NY 10016
(212) 686-9200

Michael Jay Solomon, Chairman of the Board & Chief Executive Officer; Michael N. Garin, President & Chief Operating Officer

Texscan Corporation
2960 Grand Avenue
P.O. Box 27548
Phoenix, AZ 85061
(800) 528-4066

Carl Pehlke, Chairman of the Board; Jim Luksch, President; Bob Palle, Director of Personnel

Thorn/EMI (plc)
Upper Saint Martin's Lane
London WC2 H9ED, England

United States Headquarters:
1370 Avenue of the Americas
New York, NY 10019
(212) 977-8990

Graham Powell, President; Carl Litinsky, Director of Personnel

TIE Communications, Inc.
5 Research Drive
Shelton, CT 06484
(203) 929-7373

Thomas Kelly, Jr., Chairman of the Board; William Merrit, Jr., Executive Vice-President; Alan Haas, Director of Personnel

Time, Inc.
Time & Life Building
Rockefeller Center
New York, NY 10020
(212) 586-1212

Ralph Davidson, Chairman of the Board; J. Richard Munro, President & Chief Executive Officer; Gerald Levin, Executive Vice-President, Chief Strategist & Planner; John R. Bingle, Vice-President & Personnel Director

Times Mirror Company
280 Park Avenue
New York, NY 10017
(212) 697-4070

Robert F. Erburu, President; John J. McCrory, President—
Broadcasting; Larry W. Wangberg, President—Cable Television;
Rochelle Evans, Director of Human Resources

Tribune Company
435 North Michigan Avenue
Chicago, IL 60611
(312) 222-9100

S. R. Cook, President & Chief Executive Officer; Donald Hacker,
Director of Planning & Analysis; George Veon, Director of Human Resources

Viacom International, Inc.
1211 Avenue of the Americas
New York, NY 10036
(212) 575-5175

Ralph M. Baruch, Chairman; Terrence A. Elkes, President &
Chief Executive Officer; William Seres, Director of Personnel

Warner Communications, Inc.
75 Rockefeller Plaza
New York, NY 10019
(212) 484-8000

Steven J. Ross, Chairman & Chief Executive Officer; Bert W.
Wasserman and D. F. Johnson, Office of the President; Patricia
Ross, Director of Personnel

Western Union Corporation
1 Lake Street
Upper Saddle River, NJ 07458
(201) 825-5000

Robert Flanagan, Chairman of the Board, President & Chief Executive Officer; Calvin H. Darrin, Vice-President & Controller; Stephen E. Smiszko, Vice-President & Treasurer; Robert Freifeld, Director of Professional Employment

Group W—Westinghouse Broadcasting & Cable, Inc.
888 Seventh Avenue
New York, NY 10106
(212) 307-3000

Daniel L. Ritchie, Chairman & Chief Executive Officer; William F. Baker, President TV Group & Chairman Group W Satellite Communications; John Moran, Director Human Resources

Zenith Electronics Corporation
1000 Milwaukee Avenue
Glenview, IL 60025
(312) 391-7000

Jerry K. Pearlman, Chairman, President & Chief Executive Officer; David Denton, Vice-President Human Resources

Sample Pro Forma Financial Statement for a Broadcasting Acquisition

XYZ COMMUNICATIONS OF KENTUCKY, INC.
(A Wholly Owned Subsidiary of XYZ Communications, Inc.)

PROJECTED PRO FORMA BALANCE SHEETS AND STATEMENTS OF INCOME AND CASH FLOW

1984–1989

SMITH, JONES & JENSEN

CERTIFIED PUBLIC ACCOUNTANTS

600 FOURTH AVENUE • NEW YORK, NY 10016 • TEL 212-013-1313·

Board of Directors
XYZ Communications of Kentucky, Inc.

The accompanying projected pro forma balance sheets and statements of income and cash flow of XYZ Communications of Kentucky, Inc. (a wholly owned subsidiary of XYZ Communications, Inc.) for the years 1984 to 1989 are based on information and assumptions submitted by management.

Such assumptions are described in the Notes to the projected pro forma financial statements.

Since the projections and pro forma adjustments are based upon assumptions concerning assets, liabilities, income, expenses, cash flow, and other matters which may, in actuality, be higher or lower than projected, we are precluded from expressing, and therefore do not express, an opinion on the projected pro forma balance sheets or statements of income and cash flow. Moreover, since variations of such assumptions could significantly affect the projected pro forma financial statements, we express no opinion on the reasonableness of these assumptions.

/s/ Smith, Jones & Jensen

New York, N.Y.
March 30, 1984

XYZ COMMUNICATIONS OF KENTUCKY, INC.
(A Wholly Owned Subsidiary of XYZ Communications, Inc.)

NOTES TO PROJECTED PRO FORMA BALANCE SHEETS AND STATEMENTS OF INCOME AND CASH FLOW
1984–1989

This financial forecast is based on management's assumptions concerning future events and circumstances. The assumptions disclosed herein are those which management believes are significant to the forecast or are key factors upon which the financial results of the radio station depend. Some assumptions inevitably will not materialize and unanticipated events and circumstances may occur subsequent to March 30, 1984, the date of this forecast. Therefore, the actual results achieved during the forecast period may vary materially from the forecast.

1. ASSUMPTIONS—BALANCE SHEET
    a) *Accounts Receivable*
        Accounts receivable at the end of each period is estimated at 80 days of gross revenues.

b) *Property*

Property of $1,590,000 is the fair value of the radio station's property as of the date of acquisition. (See Note 2 for depreciation policy.)

c) *Goodwill*

Goodwill represents the excess of the estimated purchase price of the station over the fair values assigned to the assets acquired. (See Note 2 for amortization policy.)

d) *Accounts Payable*

Accounts payable at the end of each period is estimated at 45 days of operating expenses.

e) *Revolving Loan*

The revolving loan of $250,000 is assumed to be borrowed and repaid each year. The interest rate is assumed to be 12½%.

f) *Long-Term Debt*

Long-term debt includes the following:

1) $1,750,000 loan from a bank bearing interest at 12½% per year with annual installments of $200,000 beginning in 1986.

2) $825,000 purchase money mortgage bearing interest at 9% per year with annual installments of $100,000 beginning in 1986.

3) $900,000 purchase money mortgage which is non-interest bearing and payable only in the event of a sale or other disposition.

2. ASSUMPTIONS—STATEMENT OF INCOME AND CASH FLOW

a) *Gross Revenues*

Gross revenues for year 1 are estimated by management based on the previous history of the station.

Gross revenues in years 1985 through 1989 are projected to increase at the rate of 10% per year.

b) *Agency Commissions*

Commissions are provided at 11% of gross revenue which includes national sales subject to a 15% agency commission and local sales not subject to an agency commission.

c) *Operating Expenses*

Operating expenses were projected at 69% of net revenues.

d) *Depreciation*

Depreciation was computed using both the straight-line and accelerated methods over the estimated useful lives of the asset.

e) *Amortization*

Amortization of goodwill was computed over a 40-year life.

f) *Management Fee*

The company is charged a management fee by XYZ Communications, Inc., its parent.

g) *Income Taxes*

Income taxes were estimated at an effective rate of 50% for Federal and local taxes. Such rate was applied to income before income taxes after adding back to such income the amortization of goodwill, which is not deductible for income tax purposes. Income taxes reflect the benefit derived from the utilization of the net operating loss carryforwards.

3. ASSUMPTIONS—OTHER

No recognition has been given to the potential income which could be derived from investing the projected cash balances in various marketable securities.

XYZ COMMUNICATIONS OF KENTUCKY, INC.

(A Wholly Owned Subsidiary of XYZ Communications, Inc.)

PROJECTED PRO FORMA BALANCE SHEETS

	1984	1985	1986	1987	1988	1989
ASSETS						
CASH	330748	538356	488150	543647	712020	930091
ACCOUNTS RECEIVABLE	436135	479748	527723	580496	638545	702400
TOTAL CURRENT ASSETS	766883	1018104	1015873	1124143	1350565	1632491
FIXED ASSETS	1590000	1590000	1590000	1590000	1590000	1590000
LESS ACCUMULATED DEPRECIATION	245167	587634	916201	1211434	1506667	1540000
NET BOOK VALUE	1344833	1002366	673799	378566	83333	50000
GOODWILL	3310000	3310000	3310000	3310000	3310000	3310000
LESS ACCUMULATED AMORTIZATION	82750	165500	248250	331000	413750	496500
NET BOOK VALUE	3227250	3144500	3061750	2979000	2896250	2813500
TOTAL ASSETS	5338966	5164970	4751422	4481709	4330148	4495991
LIABILITIES AND STOCKHOLDERS EQUITY						
ACCOUNTS PAYABLE	169275	186202	204823	225305	247835	272619
CURRENT MATURITY OF LONG-TERM DEBT:						
REVOLVING LOAN	250000	250000	250000	250000	250000	250000
BANK LOAN		200000	200000	200000	200000	200000
PURCHASE MONEY MORTGAGE—						
INTEREST BEARING		100000	100000	100000	100000	100000
INCOME TAXES PAYABLE	0	0	0	0	35532	261905
TOTAL CURRENT LIABILITIES	419275	736202	754823	775305	833367	1084524
LONG-TERM DEBT:						
BANK LOAN	1750000	1550000	1350000	1150000	950000	750000
PURCHASE MONEY MORTGAGE—						
INTEREST BEARING	825000	725000	625000	525000	425000	325000
PURCHASE MONEY MORTGAGE—						
NON-INTEREST BEARING	900000	900000	900000	900000	900000	900000
TOTAL LIABILITIES	3894275	3911202	3629823	3350305	3108367	3059524
CAPITAL STOCK	1600000	1600000	1600000	1600000	1600000	1600000
RETAINED EARNINGS	-155309	-346232	-478401	-468596	-378219	-163533
TOTAL STOCKHOLDERS EQUITY	1444691	1253768	1121599	1131404	1221781	1436467
	5338966	5164970	4751422	4481709	4330148	4495991

SEE NOTES TO PROJECTED PRO FORMA FINANCIAL STATEMENTS

XYZ COMMUNICATIONS OF KENTUCKY, INC.

(A Wholly Owned Subsidiary of XYZ Communications, Inc.)

PROJECTED PRO FORMA STATEMENTS OF INCOME AND CASH FLOW

	1984	1985	1986	1987	1988	1989	TOTALS
GROSS TIME SALES	2225800	2448380	2693218	2962540	3258794	3584673	17173405
AGENCY AND REP COMMISSIONS	235935	259528	285481	314029	345432	379975	1820380
NET REVENUES	1989865	2188852	2407737	2648511	2913362	3204698	15353025
BROADCAST EXPENSES	1373007	1510308	1661339	1827473	2010220	2211242	10593589
OPERATING INCOME	616858	678544	746398	821038	903142	993456	4759436
OTHER EXPENSES:							
DEPRECIATION	245167	342467	328567	295233	295233	33333	1540000
AMORTIZATION—GOODWILL	82750	82750	82750	82750	82750	82750	496500
INTEREST—REVOLVING LOAN	31250	31250	31250	31250	31250	31250	187500
INTEREST—OUTSIDE	293000	293000	276000	242000	208000	174000	1486000
MANAGEMENT FEE	120000	120000	160000	160000	160000	160000	880000
TOTAL OTHER EXPENSES	772167	869467	878567	811233	777233	481333	4590000
INCOME BEFORE INCOME TAXES	− 155309	− 190923	− 132169	9805	125909	512123	169436
INCOME TAXES	0	0	0	0	35532	297437	332969
NET INCOME (LOSS)	− 155309	− 190923	− 132169	9805	90377	214686	− 163533
ADD BACK (SUBTRACT):							
DEPRECIATION AND AMORTIZATION	327917	425217	411317	377983	377983	116083	2036500
SHAREHOLDER INVESTMENT	1600000						1600000
REVOLVING LOAN	250000	0	0	0	0	0	250000
BANK LOAN	1750000	0	− 200000	− 200000	− 200000	− 200000	950000
PURCHASE MONEY MORTGAGE—INTEREST BEARING	825000	0	− 100000	− 100000	− 100000	− 100000	425000
PURCHASE MONEY MORTGAGE— NON-INTEREST BEARING	900000						900000
PURCHASE PRICE OF STATION	− 4900000						− 4900000
CHANGE IN RECEIVABLES	− 436135	− 43613	− 47975	− 52773	− 58049	− 63855	− 702400
CHANGE IN PAYABLES	169275	16927	18621	20482	22530	24784	272619
CHANGE IN INCOME TAXES PAYABLE	0	0	0	0	35532	226373	261905
CHANGE IN CASH	330748	207608	− 50206	55497	168373	218071	930091
CASH—BEGINNING OF YEAR		330748	538356	488150	543647	712020	
CASH—END OF YEAR	330748	538356	488150	543647	712020	930091	930091

SEE NOTES TO PROJECTED PRO FORMA FINANCIAL STATEMENTS

Appendix D

Some Professionals You May Find It Useful to Consult

Lawyers

David R. Anderson
Wilmer, Cutler & Pickering
1666 K Street, N.W.
Washington, DC 20006
(202) 872-6000

Allen H. Arrow (E)*
Arrow Edelstein Gross &
    Asher, P.C.
919 Third Avenue
New York, NY 10022
(212) 371-7111

Jerome S. Boros
Fly, Shuebruk, Gaguine,
    Boros, Schulkind and
    Braum
45 Rockefeller Plaza
New York, NY 10111
(212) 247-3040

George R. Borsari, Jr.
1830 Jefferson Place, N.W.
Washington, DC 20036
(202) 296-8900

Linda Cinciotta
Arent, Fox, Kintner, Plotkin
    & Kahn
1050 Connecticut Avenue,
    N.W.
Washington, DC 20036
(202) 857-6029

Harold Farrow
Farrow, Schildhause, Wilson
    & Rains
401 Grand Avenue
P.O. Box 2290
Oakland, CA 94621
(415) 839-4500

Stuart Feldstein
Fleischman & Walsh
1725 N Street, N.W.
Washington, DC 20036
(202) 466-6250

Leonard Franklin (E)
Michael Rudell (E)
Franklin, Weinrib, Rudell &
    Vassallo, P.C.
950 Third Avenue
New York, NY 10022
(212) 935-5500

*(E) denotes lawyers essentially engaged in entertainment law.

Henry Goldberg
Goldberg & Spector
1920 N Street, N.W.
Washington, DC 20036
(202) 429-4900

James C. Goodale
Devevoise & Plimpton
299 Park Avenue
New York, NY 10171
(212) 909-6000

Elizabeth Granville (C)*
Kronish, Lieb, Shainswit,
     Weiner & Hellman
1345 Avenue of the Americas
New York, NY 10105
(212) 841-6290

Morton I. Hamburg
Sheber Pomerantz Slotnik &
     Hamburg
445 Park Avenue
New York, NY 10022
(212) 751-6700

Ronald S. Konecky (E)
Kay, Collyer & Boose
One Dag Hammarskjold
     Plaza
New York, NY 10017
(212) 940-8200

Erwin G. Krasnow
Verner Liipfert Bernhard &
     McPherson
Suite 1000
1660 L Street, N.W.
Washington, DC 20036
(202) 452-7400

Robert Lasky (E)
598 Madison Avenue
New York, NY 10022
(212) 486-2727

Robert H. Montgomery, Jr. (E)
Paul, Weiss, Rifkind,
     Wharton & Garrison
345 Park Avenue
New York, NY 10154
(212) 644-8000

Preston Moore
Morrison & Foerster
One Market Plaza
Spear Street Tower
San Francisco, CA 94105
(415) 777-6000

Alan Y. Naftalin
Koteen & Naftalin
1150 Connecticut Avenue
Washington, DC 20036
(202) 467-5700

W. Theodore Pierson, Jr.
Pierson, Ball & Dowd
Ring Building
1200 18th Street, N.W.
Washington, DC 20036
(202) 331-8566

Ernest T. Sanchez
Liberman, Sanchez &
     Bentley
Suite 200
2000 L Street, N.W.
Washington, DC 20036
(202) 887-5068

*(C) denotes lawyers essentially engaged in copyright and trade mark law.

Bruce P. Saypol
Wood, Lucksinger & Epstein
Suite 500
2000 M Street, N.W.
Washington, DC 20036
(202) 223-6611

Sol Schildhause
Farrow, Schildhause, Wilson
    & Rains
1710 Rhode Island Avenue
Washington, DC 20036
(202) 822-8300

Richard M. Schmidt, Jr.
Cohn & Marks
1333 New Hampshire
    Avenue, N.W.
Washington, DC 20036
(202) 293-3860

Andrew Jay Schwartzman
Media Access Project
1609 Connecticut Avenue,
    N.W.
Washington, DC 20009
(202) 232-4300

Stuart A. Shorenstein
Friedman, Leeds &
    Shorenstein
655 Third Avenue
New York, NY 10017
(212) 867-9760

Melvin Simensky (E)
Gerstein, Savage, Kaplowitz
    & Simensky
150 East 58th Street
New York, NY 10155
(212) 752-9700

Frank Weissberg (E)
Alan J. Hartnick (C)
Colton, Weissberg, Hartnick,
    Yamin & Sheresky
501 Park Avenue
New York, NY 10022
(212) 371-4350

Donald P. Zeifang
Baker & Hostetler
818 Connecticut Avenue
Washington, DC 20006
(202) 861-1500

Roger L. Zissu (C)
Cowan, Liebowitz & Latman
605 Third Avenue
New York, NY 10158
(212) 986-6272

Broadcast and Cable Television Brokers

James W. Blackburn, Jr.
Joseph M. Sitrick
Blackburn & Company, Inc.
1111 19th Street, N.W.
Washington, DC 20036
(202) 331-9270

Broadcast Properties West Inc.
221 First Avenue West
Seattle, WA 98119
(206) 283-2656

Business Brokers Associates
4611 Hixson Pike
Chattanooga, TN 37343
(615) 756-7635

Chapman Associates
1835 Savoy Drive
Atlanta, GA 30341
(404) 458-9226

R. C. Crisler & Co. Inc.
580 Walnut Street
Cincinnati, OH 45202
(513) 381-7775

Gammon & Ninowski Media
    Brokers, Inc.
1925 K Street, N.W.
Washington, DC 20006
(202) 861-0960

Horton & Associates
Box 948
Elmira, NY 14902
(607) 733-7138

Bob Kimel's New England
    Media, Inc.
8 Driscoll Drive
St. Albans, VT 05478
(802) 524-5963

H. B. LaRue, Media Broker
44 Montgomery Street
Suite 500
San Francisco, CA 91404
(415) 434-1750
500 East 77th Street
New York, NY 10162
(212) 288-0737

Robert O. Mahlman, Inc.
One Stone Place
Bronxville, NY 10708
(914) 779-7003

Cecil L. Richards Inc.
7700 Leesburg Pike
Suite 408
Falls Church, VA 22043
(703) 821-2552

Robert W. Rounsaville &
    Associates
3104 Shadowlawn Avenue N.E.
Atlanta, GA 30305
(404) 261-3000

Howard E. Stark
575 Madison Avenue
New York, NY 10022
(212) 355-0405

Broadcast and Cable Engineers

Julius Cohen
Cohen & Dippell P.C.
1015 15th Street, N.W.
Washington, DC 20005
(202) 883-0111

Malarkey, Taylor & Associates
1301 Pennsylvania Avenue, N.W.
Washington, DC 20004
(202) 737-2300

Wallace E. Johnson
Moffet, Larson & Johnson
1925 North Lynn Street
Arlington, VA 22209
(703) 841-0500

Dana Puopolo
Puopolo Communications, Inc.
P.O. Box 72
South Attleborough, MA
    02703
(401) 723-3775

A. D. Ring & Associates
1140 19th Street, N.W.
Washington, DC 20036
(202) 223-6700

Neil M. Smith
Smith & Powstenko
2000 N Street, N.W.
Washington, DC 20036
(202) 293-7742

Broadcasting Sales Representatives

Avery-Knodel Television
437 Madison Avenue
New York, NY 10022
(212) 421-5600

John Blair & Company
1290 Avenue of the Americas
New York, NY 10104
(212) 603-5000

Jack Bolton Associates
Suite 417
3384 Peachtree Road, N.E.
Atlanta, GA 30326
(404) 237-1577

Cabellero Spanish Media Inc.
310 Madison Avenue
New York, NY 10017
(212) 972-1019

CBS Radio Spot Sales
51 West 52nd Street
New York, NY 10019
(212) 975-4321

CBS Television Stations
    National Sales
51 West 52nd Street
New York, NY 10019
(212) 975-4321

The Christal Company
919 Third Avenue
New York, NY 10022
(212) 688-4414

Eastman Cablerep
1 Rockefeller Plaza
New York, NY 10020
(212) 581-9771

Eastman Radio Inc.
1 Rockefeller Plaza
New York, NY 10020
(212) 581-0800

Group W Television Sales
    Inc.
90 Park Avenue
New York, NY 10016
(212) 883-6100

Katz Radio
Katz Television
One Dag Hammarskjold
    Plaza
New York, NY 10017
(212) 572-5500

Major Market Radio Inc.
415 Madison Avenue
New York, NY 10017
(212) 355-1700

Jack Masla & Company Inc.
41 East 42nd Street
New York, NY 10017
(212) 490-3760

McGavren-Guild Inc.
154 East 46th Street
New York, NY 10017
(212) 599-7394

NBC Spot Television Sales
30 Rockefeller Plaza
New York, NY 10020
(212) 664-4444

Peter, Griffin, Woodward Inc.
645 Fifth Avenue
New York, NY 10022
(212) 826-6000

Petry Television Inc.
3 East 54th Street
New York, NY 10022
(212) 688-0200

Selcom Radio
521 Fifth Avenue
New York, NY 10017
(212) 490-6620

SIN Television Network
460 West 42nd Street
New York, NY 10036
(212) 953-7500

Southern Spot Sales Inc.
Box 18006
Raleigh, NC 27619
(919) 782-0896

TeleRep Inc.
875 Third Avenue
New York, NY 10022
(212) 759-8787

Glossary of Telecommunications Terms

affiliate—an independently owned television or radio station that contracts with a broadcasting network to carry network programs. Generally, the affiliate is paid by the network for carrying such programs and advertising.

amplitude modulation (AM)—variations of the amplitude of a radio wave in accordance with the sound being broadcast. Signal reception occurs either via ground waves or sky waves which are bounced off the ionosphere and reflected back to Earth. AM signals are not subject to topographic obstructions. The frequencies on which AM radio stations broadcast are measured in kilohertz. In the United States, these frequencies are assigned by the Federal Communications Commission.

amplitude—the relative height and depth of the peaks and troughs of sound waves. Where two sound waves are traveling at the same frequency, the one with the greater amplitude will be perceived as louder.

basic cable—the transmission of television signals by wire (cables) or fiber optics using laser technology. This provides improved reception and increased availability of retransmitted local and long distance stations. Generally, "basic" is the minimum cable programming service provided by a cable television company. It often consists of public access and information channels, local television stations, long distance television signals, and certain satellite-delivered programming.

cablecasting—both the retransmission of programming from various sources, through a cable television system, and the transmission of programming that originates at the cable television system operator's headquarters or studios.

community antenna television (CATV)—direct service by wire transmissions from a common antenna that delivers high quality signals. The system was originally developed for areas with poor to nonexistent television reception, but it is now being sold in major metropolitan areas. The CATV service is provided on a fee basis, by an independent company. See also *satellite master antenna television* (SMATV).

capacitance electronic disc (C.E.D.)—a consumer videodisc system brought onto the market, and now terminated, by RCA. The system uses a diamond stylus guided by microscopic grooves on the disc. The video signal is transmitted through changes in the capacitance between stylus and disc. The videodisc is not compatible with other disc formats.

cellular radio-telephone systems—mobile telephone systems in which many antennae are placed within a grid of cells over a particular geographic area, generally one metropolitan market, with one antenna per cell. Several channels are transmitted by each antenna. A computerized system routes calls and locates available channels as a caller drives from one cell to the next. The system offers unlimited subscriber capacity and long-range roaming capability. Where the non-cellular mobile telephone could place approximately 25 calls at one time, the cellular system can place as many as 1200 calls at once.

communications satellite (geostationary)—an electronic space vehicle in orbit 22,300 miles above the equator. At such height, the satellite travels at the same rate of speed as the Earth, thus giving the appearance of being stationary in one place above the earth. The satellite receives electronic signals from Earth and retransmits them back to Earth station receivers or "dishes." The signals contain programming and other data.

direct broadcast satellite (DBS)—a high-powered satellite authorized to broadcast directly into homes that are equipped with appropriate antennae or "dishes."

earth station—the electronic equipment used in conjunction with the satellite antenna or "dish." Receives, processes, and transmits radio frequency signals to and from satellites.

Federal Communications Commission (FCC)—Federal agency established by Congress in 1934 to regulate radio and television broadcasting and to assign broadcasting licenses. The FCC represents the public interest.

fiber optics—a technology allowing transmission of information in the form of light through special glass fibers. The fiber optics cables can carry a variety of materials, television broadcasting and other information, on the same lines. The fibers are less than one-fifth of an inch in diameter and are resistant to electromagnetic interference. They can also transmit laser light beams over long distances.

franchise—a contract or agreement between a CATV company and the franchise granting authority (usually a local government or municipality). The agreement specifies the rights and duties of the parties with regard to the construction and operation of the system for the agreed-upon area.

frequency allocation—the FCC assignments of specific frequencies to radio and television broadcasters.

frequency modulation (FM)—the variations in the frequencies of a radio wave in accordance with the sound being broadcast. FM is a high-fidelity, line-of-sight beam and is impeded by topographic obstructions, but not by atmospheric interference. The frequencies on which FM radio stations broadcast are measured in megahertz. In the United States, these frequencies are assigned by the Federal Communications Commission.

group-owned station—a radio or television station licensed to a corporation owning two or more stations.

headend—the CATV system control center. It processes signals and transmits them to subscribers.

hertz—a unit of frequency equal to one cycle per second. AM frequencies are measured in kilohertz, and FM frequencies are measured in megahertz. Named after Heinrich Hertz, their discoverer.

high definition television (HDTV)—a system that transmits television signals of more lines per frame than the current U.S. standard of 525 lines per frame. The result is a vastly improved picture.

independent station—according to the FCC, an independent station is a commercial television station that carries a maximum of ten hours of prime-time network programming per week.

laser disc—a video playback system developed by Philips. A low-powered laser beam reflects off the disc and reads the information, which is encoded in a series of microscopic pits on the disc. In an audio record, the walls of grooves are modulated with an audio signal. The laser disc tracks have finer modulation, and the spacing is sixty times smaller than the audio record. This technology permits random access and interactivity, particularly when combined with a computer.

low power television (LPTV)—an FCC proposal that would license television stations in the United States to operate at low power serving a fifteen-to-twenty-mile radius. These stations would not be required to observe the minimum geographical separation standards established for full power stations that are on the same channel.

master antenna television (MATV)—a system that delivers pay programming to multi-unit dwellings. Tenants connect their sets to an outlet connected to the MATV on the building's roof.

microwave—electromagnetic waves of high radio frequencies beginning at 1000 MHz, but small amplitude. Used to transmit television signals over long distances.

mobile radio-telephones—a technology using one high powered antenna to transmit several dozen radio channels across a fifty-mile radius. Since only one person at a time can have access to a channel, up to 90 percent of all dialing efforts may result in a busy signal.

multichannel television sound (MTS)—a system approved in March 1984 by the FCC which will use amplitude modulation (AM) of sound subcarriers.

multiple system operator (MSO)—a corporation that owns two or more cable systems.

multipoint distribution service (MDS)—the delivery of a pay-per-channel television service, using omnidirectional line-of-sight microwave signals. This is becoming less popular as cable and multichannel MDS gain audiences.

network programming—television programming produced by the network and carried by subscribers (network affiliates). The broadcast system is linked by wire, microwave, or satellite so that all affiliates can carry the same programs simultaneously.

pay cable television—a cable television service that offers special movies, sports, and/or other programming to subscribers for a monthly fee in addition to the basic cable charge. The company contracts with a cable television system operator for a given geographic area and then offers its service to subscribers.

radio frequency (RF)—a portion of the electromagnetic frequency spectrum used for transmission of radio and television signals either through space or wires. Radio frequency signals travel at the speed of light and are measured in hertz.

rating—the percentage of homes out of the total potential audience, in a given geographic area, viewing a particular program. Ratings are relied upon in selling radio and television time to advertisers.

satellite master antenna television (SMATV)—a system installed in multi-unit dwellings to distribute pay services but which avoids local franchising requirements as long as no streets are crossed or involved. Satellite signals are received through a dish on top of the building. Installation costs are a fraction of the cost of cable.

satellite station—a television station that sends signals to areas beyond the coverage area of the parent station. No programming originates from the satellite station.

subscription television (STV)—a broadcast system using UHF or VHF signals to transmit programs from the television station to subscribers within a certain geographic area. Subscribers are equipped with a decoding device that unscrambles the signals.

superstation—a television station that transmits its signals to a satellite and then to receivers owned by cable television systems, enabling the cable subscribers to view programming of local stations in distant cities. An example is WTBS-TV, owned by Turner Broadcasting Corporation, in Atlanta, Georgia, which sells its signal to cable systems throughout the United States.

syndicated programs—programs offered for sale by, and sometimes produced by, independent companies to stations and advertisers who do not want or cannot afford network programming.

teletext—a service displaying information on home televisions. The information is catalogued and can be accessed through a keyboard. Supplies a range of information, from news, sports results, and stock market quotations to traffic and shopping data.

translator—a television station that rebroadcasts signals from another station. No programming originates from the translator stations.

transponder—a transmitter and receiver located on an Earth-orbiting satellite. The satellite picks up signals from Earth, then translates them to a new frequency, amplifies them, and retransmits them back to Earth. Generally, there are twenty-four transponders on each domestic communications satellite. Companies seeking to use a particular transponder for transmitting programs generally lease one or more transponders for a period of time from the company placing the satellite in space, and pay established monthly rents. In the United States, the satellite is actually placed in space by the

National Aeronautical and Space Administration (NASA) under contract.

ultrahigh frequency (UHF)—from 300 to 3000 megahertz. Transmission costs are higher than VHF.

videocassette recorder (VCR)—a video recording device that uses a helical scan technique employing moving video heads attached to a cylinder or drum that rotates at 1800 rpm. This technique permits slow-moving half-inch tape to hold high frequencies. The VCR can record programs off the air and can play back prerecorded tapes.

very high frequency (VHF)—television channels 2 through 13, broadcast at 30 to 300 megahertz. Uses less energy than UHF.

videotape—a helical scan recording medium used for television that consists of polyester carbon and oxide bases in standard two-inch size for television broadcasting, one-inch for cable, and three-quarter-inch and half-inch sizes for video recording technology.

Index